AN ILLUSTRATED GUIDE TO

MODERN
FIGHTER
COMBAT

AN ILLUSTRATED GUIDE TO

MODERN
FIGHTER
COMBAT

PRENTICE HALL PRESS
New York   London   Toronto   Sydney   Tokyo

Mike Spick

A Salamander Book

All correspondence concerning the content
of this book should be addressed to
Salamander Books Ltd.,
52 Bedford Row,
London WC1R 4LR,
United Kingdom.

An Arco Military Book

Pubished in 1987 by Prentice Hall Press
A Division of Simon & Schuster, Inc.
Gulf + Western Building
One Gulf + Western Plaza
New York, NY 10023

PRENTICE HALL PRESS is a trademark of
Simon & Schuster, Inc.

Originally published by
Salamander Books Ltd., London

Library of Congress Cataloging-in-Publication Data

Spick, Mike.
  An illustrated guide to modern fighter combat.

  (An Arco Military Book)
  1. Fighter plane combat. I. Title. II. Series.
UG700.S634   1987   358.4'3   86-25560
ISBN 0-13-451055-0

10 9 8 7 6 5 4 3 2

Contents

Credits

Author: Mike Spick is the author of several works on modern combat aircraft and the tactics of air warfare, including Salamander's *Modern Air Combat* and *Modern Fighting Helicopters* (both with Bill Gunston) and Fact Files on the F-4 Phantom (with Doug Richardson), F-14 Tomcat and F/A-18 Hornet.

Editor Bernard Fitzsimons
Art Editor Mark Holt
Designed by TIGA

Diagrams: TIGA

Typeset by The Old Mill, London
Color reproduction by Melbourne Graphics
Printed in Belgium by Proost International Book Production, Turnhout

Acknowledgements: The pubishers are grateful to all the companies and other organizations who supplied photographs for use in this book.

Weapon Systems

It is the fighter aircraft that catches our attention as it trails thunder across the sky, often sleek and beautiful, occasionally rather ugly, but invariably epitomising power and menace. Neither the featureless air superiority grey with low-visibility markings, nor the spectacular colour schemes often sported by prototypes at international air shows detract from this impression. Whether dressed in working garb or carrying the full cosmetic treatment, the fighter is designed to kill, and looks it. Yet it is in fact no more than a carrier of weapons, and its sole function is to get those weapons into a position from which they can be used with the best possible chance of success, at the same time avoiding positions where an opponent's weapons can be successfully launched at it.

The fighter as we see it is a machine that flies, a graceful piece of metal sculpture, an example of modern art. Yet flight is not the only measure of its capability. The sleek and simple-looking missiles that hang beneath the wings betray its purpose, to blast other aircraft out of the skies, in a welter of flame and torn metal, while buried deep in its vitals are the mysterious black boxes which enable it to carry out its mission. These, the avionics, are as much a part of the machine that bears them as the wings or the tail. The fighter is a weapon system, a tightly knit package of metal and power, and of electronics that enable it to detect foes far beyond the range of human eyesight in bad weather and darkness, and that warn it of threats from air or ground and enable it to take action to confound them.

Below: Pre-flight checks for an F-15D pilot. The man in the cockpit is the fighter's own intelligence, dependent on its life support systems for protection from a hostile environment.

The fighter carries its own intelligence in the form of the pilot, who sits in his small cockpit, protected from a hostile environment by the on-board life support systems. The progress made in avionic systems over the past two decades means that he is at the centre of an information explosion; the aircraft systems gather more information than he can ever hope to assimilate. More clever systems filter and process this information, and present the pilot with just what he needs to know for the accomplishment of the current part of the mission. Other systems can call his attention to emergency situations.

Below: Advances in avionics mean the F/A-18 pilot has information filtered, processed and presented to him in a form that is much more readily assimilable.

The information is presented in amounts that the pilot can (just) handle. The HOTAS ensures that everything that he needs in critical flight situations such as combat, or landing, is directly under his hands, although this calls for a high degree of manual dexterity, while essential information can be flashed up onto the HUD to obviate the need to look down into the cockpit at a crucial point in the proceedings.

The remarkable achievement that is the modern fighter stands at the pinnacle of uncountable man-years of effort, research and bitter combat experience. To the designers it is something else, the outcome of a series of hard-fought compromises, of theories accepted and rejected. It is also an exercise in packaging; the art of fitting things into a finite and often inconveniently shaped volume.

The sole purpose of the fighter is to bring weapons to bear on an opponent and to avoid having them brought to bear on oneself. The design of the fighter, and by this we mean not only the airframe, but the powerplant and the black boxes that go to make up the whole, is therefore conditioned by the weapons that it will carry, and also by the weapons by which it will be opposed.

The weapons of air combat fall into two basic categories, the gun and the homing rocket. The latter, more usually known as the guided missile, is by far the more important, and must be dealt with in some detail.

Manoeuvring missiles have various types of guidance systems. The specific type will determine the range and angle at which they can be launched, and the radar cross-section or flight profile of targets against which they can be used. The one thing that they all have in common is that the manufacturer's brochure figures are fairly meaningless when it comes to combat. If we are told that the maximum speed of the missile is Mach 4,

Below: On his left the Eagle pilot has radar, fuel and other controls as well as the throttles with their multi-function switches.

Right: McDonnell Douglas pioneered the HOTAS approach on the F-15 and refined it for use on the F/A-18 Hornet.

its time of flight is 60 seconds, and its range is 30nm (56km), then we may be fairly sure that if launched at a suitable speed from high altitude it will achieve all these conditions. We may also be told that it can pull a loading of 30g in a turn, and it probably can.

Taken at face value, this makes the fate of an enemy fighter flying at Mach 1, 20nm (37km) away and able to pull a mere 9g in the turn, seem fairly predictable. In practice, speed, range and g all vary enormously with the altitude of both launching fighter and target, while relative speeds,

F/A-18 HOTAS controls

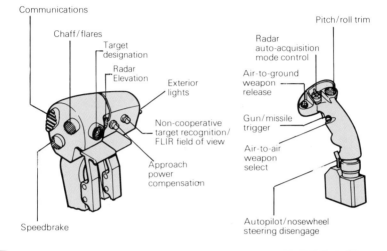

Communications
Chaff/flares
Target designation
Radar Elevation
Exterior lights
Non-cooperative target recognition/ FLIR field of view
Approach power compensation
Speedbrake

Pitch/roll trim
Radar auto-acquisition mode control
Air-to-ground weapon release
Gun/missile trigger
Air-to-air weapon select
Autopilot/nosewheel steering disengage

especially the speed of the launching fighter, and target aspect all play a great part in modifying the brochure figures, as does the sensitivity of the missile seeker head.

To make some rough generalisations, the brochure speed and range of the missile are correct for a launching fighter speed of about Mach 1 at around 40,000ft (12,200m). The higher the altitude, the lower the atmospheric pressure on the back chamber of the rocket motor and the greater the thrust obtainable. At the same time, the attenuated atmosphere causes less drag and enables the missile to fly further than it could lower down.

Most missiles have a short motor burn time. In the case of short-range (visual range) weapons, this is only a couple of seconds, during which the missile is accelerated to its maximum velocity. Medium-range (beyond visual range, or BVR) weapons normally have a sustainer motor which takes over after the initial acceleration phase and lasts for a few more seconds before it too burns out from lack of fuel.

The time that a missile is actually at its stated maximum speed is therefore very brief for a visual range weapon, and only a few seconds for a BVR missile. After motor burnout the missile is just coasting along, flying slower and slower until control is lost and it falls. In the latter stages of flight it can be outrun by a fast moving fighter.

What, then, of a missile that can manoeuvre at 30g? It can be assumed that missile manoeuvrability maximises at the point when its velocity is highest and its weight is lowest, which occurs at about the point of motor burnout. Thereafter its energy is slowly being depleted, which reduces its manoeuvre capability, while any manoeuvring will further deplete its energy.

In any case, what is the value of a 30g turn? At a velocity of Mach 4 at or above the tropopause, it gives a turn radius of about 15,600ft (4,750m) and a turn rate of 14.26°/sec. This is nothing very special, and can be bettered by almost any fighter flying at subsonic speeds, provided the pilot knows that the missile is on its way, and from what direction, and exactly when to break to avoid it.

So far we have been looking at a missile at high altitude. What happens lower down, where less thrust is obtainable and the drag is greater due

Below: Missile range at sea level may be as little as one-third the value for 40,000ft, where the thinner air causes less drag and rocket thrust is increased.

Launch envelope variation by altitude

Approximate maximum aerodynamic missile range

Launch envelope: non-manoeuvring target

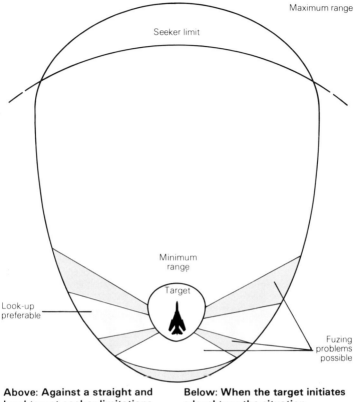

Above: Against a straight and level target seeker limitations and fuzing problems may reduce the maximum range of a semi-active radar homing missile.

Below: When the target initiates a level turn the situation becomes much more complicated and the envelope is greatly restricted.

Launch envelope: manoeuvring target

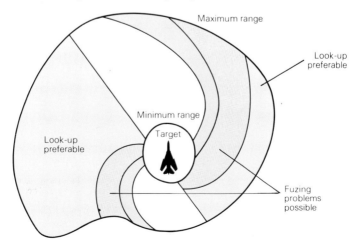

to the denser atmosphere? The short answer is that the missile is slower and does not fly so far. Exact figures are variable according to the missile type, but velocities may fall by as much as one third, say from Mach 3 to Mach 2, while range reduces typically by two thirds, for example from 9nm (17km) to 3nm (6km). This is a vast difference, and it alters the tactical situation considerably. Short-range missiles are the worst affected, as the sustainer motor of the medium range type, with its longer burn, gives some extra distance.

So far we have dealt with absolute missile range, or perhaps it should be called travel distance. What really counts in air combat is launch range, and this is a different matter entirely.

A missile is in flight for a finite time, and during that time the target is moving. Missile range is a static measurement while launch range is a dynamic one. It is moderated by the closure, the speed at which the launching fighter and the target are coming together, or the negative closure, the rate at which the distance between them is increasing.

The accompanying diagram shows a very fast missile launched from a fighter travelling at Mach 1 against a co-speed, co-altitude target approaching from head-on, and another going directly away. In the head-on attack, the launch range is 25nm (46km), and in the attack from astern it is down to 15nm (37km) with a flight time of 30 seconds. This is a

Below: Head-on and astern attack ranges for a missile with an effective flight time of 30 seconds; both aircraft are assumed to be at Mach 1 and 25,000ft (7,620m). In both cases the missile covers 20nm, but the target heading varies the range at which it can be engaged from 15nm astern to 25nm head-on.

Semi-active radar homing missile attack ranges

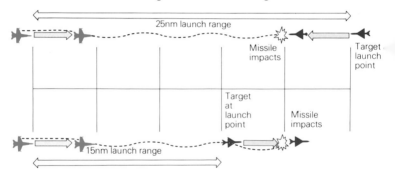

Below: The Hughes AIM-54A Phoenix carried by the F-14 Tomcat is the only really long-range air-to-air missile. In one of the most dramatic trials of its capabilities, staged in April 1973, a Backfire-type target was engaged at a range of over 100nm, the missile covering 72.5nm and reaching 103,500ft (31,500m) before passing within lethal range of the simulated bomber.

Phoenix long-range capability

F-14 at Mach 1.5 and 14,000ft acquires target at 132nm range

Phoenix launched at 110nm range

Position of F-14 at missile impact assuming course and speed held

Peak altitude of missile 103,500ft

simplistic representation; a rough rule of thumb would be to add 50 per cent to the static range for the head-on attack under these conditions, and deduct about one third from the static range for the attack from astern.

The true long-range missile is a rare bird in the air combat world, as it must be large and expensive, and needs a large and specialised fighter to carry it. The current world record for a long-range interception is held by the Hughes AIM-54 Phoenix, an example of which was launched from a Tomcat of the US Navy in April 1973 at a high-altitude supersonic target approaching from head-on, and passed within lethal distance.

The launch was made from a distance of 110nm (204km), although the distance travelled by the Phoenix was just 72½nm (134km). This remarkable distance was only achieved by pre-programming the big missile into a high trajectory, peaking

Launch envelope variation by closure speed

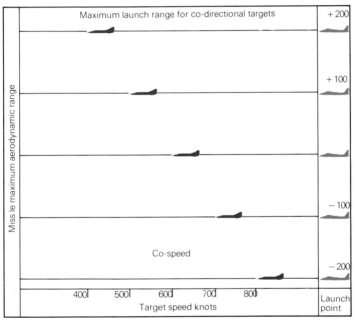

Above: Given a missile with an effective flight time of 30 seconds and an average speed of Mach 2, an attacker travelling at 600kt engaging a target from astern will find his launch enve- lope severely restricted if the target has a speed advantage. Even against a slower target he will not be able to exploit the missile's theoretical maximum aerodynamic range.

BQM-34E augmented to simulate radar cross-section of Backfire at Mach 1.5 and 50,000ft

Missile impacts 72.5nm from launch point

Position of target at Phoenix launch

at 103,500ft (31,500m), thus giving it positional energy in the form of height which could be traded for kinetic energy in the form of speed to maintain its manoeuvre capability at the end of its run. Phoenix, which has a brochure speed of Mach 4, averaged less than Mach 3 during this interception.

The final problem common to all missiles, although this is beginning to be overcome, is fuzing. If a direct hit is scored, all well and good, but often the launch results in a very near miss, and a proximity fuze is needed to make the kill. The difficulty arises from the vast disparity of interception angles that are possible, with attendant speed variations. Obviously, there needs to be a fractional delay between the proximity fuze detecting the target and the detonation of the warhead.

Front and rear quarter attacks tend to run the length of the target aircraft, and provided that the missile explodes somewhere along this line, lethal damage is very likely to occur. Missiles approaching from the beam, on the other hand, stand a good chance of overshooting before detonating, causing no damage, and the same thing can happen in reverse in a maximum range astern attack, when the built-in fuze delay detonates the warhead short of its target, in which case limited damage may be caused, but not the maximum. Finally, the fuze must not be detonated by the launching aircraft; a built-in delay while the fuze arms

Below: After launch from an F-16 (top), an Amraam missile homes on and smashes through a QF-102 target, which crashes in flames.

18

itself is obligatory, but causes a minimum range barrier under which it will not work. Again, this problem is gradually being overcome, but a decade ago the minimum range was of the order of half a mile.

Three types of guidance system are commonly used for AAMs. These are active radar homing, semi-active radar homing (SARH), and infra-red (IR) homing. Active radar homing demands a very large missile if it is to achieve a worthwhile range with its radar, and the carriage of such a weapon would considerably reduce the aircraft's performance. It would also be expensive to the point of being unaffordable by all but the richest nations.

A compromise solution adopted by Phoenix, and also by AIM-120 Amraam and the French MICA, is to use some form of midcourse guidance to take the missile to within a

Amraam operation

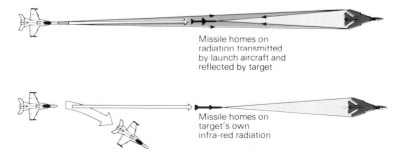

Above: Amraam allows the launch fighter to manoeuvre after launch: missile guidance is inertial, with data-link update and terminal active radar homing.

Below: Amraam (top), Sparrow (centre) and Sidewinder homing methods. The second requires illumination by the launch fighter until the missile impacts.

Missile homing methods

Inertial system updated, then missile tracks target with own radar

Missile homes on radiation transmitted by launch aircraft and reflected by target

Missile homes on target's own infra-red radiation

few miles of the target, then use the active radar for the terminal homing phase, of say, 5-10nm (9-18km). The midcourse guidance can either be preset inertial, or an update via data link from the launching fighter's radar.

SARH is rather simpler and cheaper than active radar homing: it requires the fighter to illuminate the target with its own on-board radar, enabling the missile to home on the reflected energy. This has an inbuilt disadvantage in that it requires illumination throughout the flight of the missile, making the launching fighter predictable, and therefore vulnerable, for far too long. Moreover, if it makes any but the mildest change of course, the radar lock will be broken and the missile will go ballistic.

Another failing of SARH, which is also the reason why active radar homing has a short range in practice, is that the sensitivity of the seeker head is limited. A small missile can naturally carry only a small antenna, and this severely restricts the radar returns that can be detected. This is further modified by the radar reflectivity of the target, which controls the amount of energy that is reflected back.

Sparrow is one of the most common SARH missiles in use today, and its aerodynamic range is stated as something in excess of 25nm (46km). Against a head-on target this could be increased considerably, provided

Below: A direct hit by a Sparrow means a certain kill, but near misses pose fuzing problems.

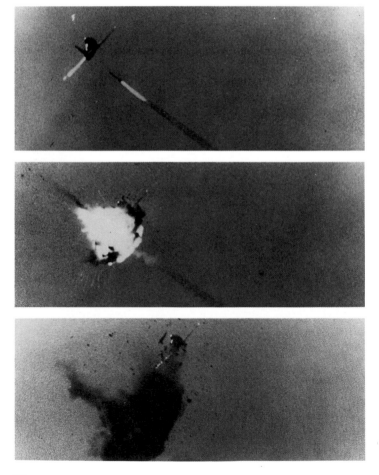

that the seeker head can detect the reflected emissions. Against a small target, such as the MiG-21 at the frontal aspect, detection and therefore homing range of the AIM-7E Sparrow is only some 7-8nm (13-15km). SARH also has problems against lower flying targets from some aspects, notably the beam, and against co-speed targets from astern.

The simplest of all missile guidance systems is IR, which homes on heat emissions from the target and permits true fire-and-forget weapons, enabling the launching fighter to manoeuvre freely immediately after launch. IR is generally used in short-range missiles; its ability to track a target attenuates rapidly with distance, and if it could be used at BVR ranges, great care would need to be taken that no friendly aircraft was anywhere near the line of flight, as it lacks discrimination, and can easily switch targets. Possibly the most serious failing of IR homing is that it becomes ineffective in cloud or heavy rain, which makes it less suitable for European conditions than it would be in, say, the Middle East.

Aircraft target signatures

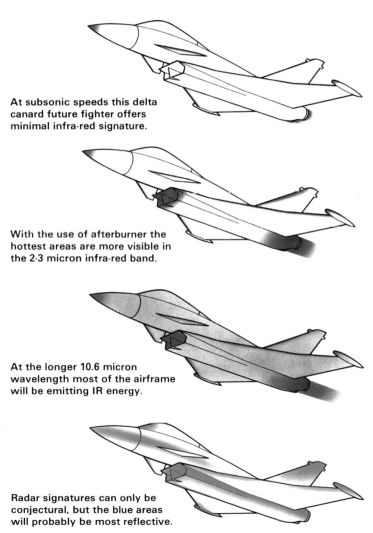

At subsonic speeds this delta canard future fighter offers minimal infra-red signature.

With the use of afterburner the hottest areas are more visible in the 2-3 micron infra-red band.

At the longer 10.6 micron wavelength most of the airframe will be emitting IR energy.

Radar signatures can only be conjectural, but the blue areas will probably be most reflective.

The final weapon in air combat is the gun. This is usually somewhere between 20mm and 30mm in calibre, and rates of fire vary between 1,200 and 6,000 rounds per minute. Either one or two are normally carried, and it is regarded mainly as a back-up weapon that is instantly available, and usable within minimum missile range, and provides a means of defence when all missiles are expended. Theoretical ranges vary, but the maximum distance at which the average pilot can be expected to hit anything is about 1,500ft (450m).

Air-to-air gunnery is a book on its own, but basically there are two types of shot; the tracking shot, with the firer following the target with the gunsight, and the snapshot, a shot of opportunity taken when tracking is not possible.

Given the weapons, the problem then arises of how best to employ them. Historically, the dominant factor in air fighting has been surprise, and roughly four out of every five pilots shot down failed to see their assailants until it was too late, if at all. (This is, of course, in fighter versus fighter combat; fighters versus a bomber formation in broad daylight are something else). Modern tech-

nology notwithstanding, there is no reason to assume that anything has changed to invalidate the surprise factor, and the crucial element in achieving surprise is detecting the enemy first. Whoever sees his opponent first gains the initiative, and an opponent on the defensive is half way to being beaten.

There are several ways of detecting the enemy. Radar is the most widely used, and takes various forms: the fighter will carry a radar and use it to actively search for the enemy; ground-based radars linked to a fighter control system can play an important part; and airborne radar in the form of airborne early warning and control AEW&C aircraft is able to control and monitor vast areas of airspace, assess threats, and, in contact with both ground control and airborne fighters via data link, exert a tremendous influence on events.

To be introduced before very long is the Joint Tactical Information Distribution System, or JTIDS. This will feed essential information directly into the avionics and radar systems

Below: The JTIDS display in an F-15 cockpit shows friendly, hostile and unidentified aircraft.

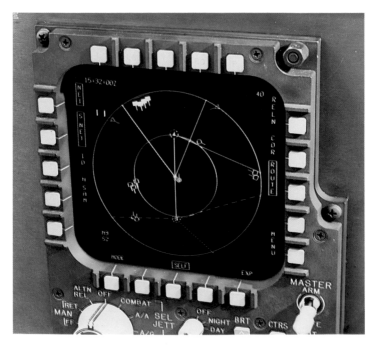

Datalink application

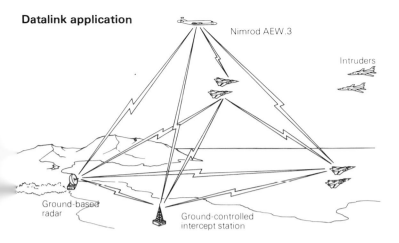

Nimrod AEW.3

Intruders

Ground-based radar

Ground-controlled intercept station

Above: How pairs of Tornado interceptors will apply information from AEW&C aircraft and ground radar stations.

Below: Tornado ADV and Nimrod AEW.3, the combination of fighter and AEW&C aircraft designed to defend UK airspace.

of fighters, enabling them to fly much of a mission with their on-board radars on standby. The advantage of this is that radar is an emission and it is detectable by the other side. When the radar is used only for brief periods, it greatly lessens the chance of detection by the enemy and increases the probability of attaining surprise. It also promises to solve the vexed problem of BVR identification.

A modern fighter radar will be multi-mode. Some of its modes will be concerned with navigation, the delivery of air-to-ground ordnance and terrain avoidance, but others will be for air combat. The Hughes APG-65, currently flying in the F-18, selected for the updating of Luftwaffe F-4Fs and a contender for the Eurofighter, has a typical selection of air to air modes coupled with high resolution alpha-numeric displays. The modes are:

Velocity search, used for long range detection against rapidly closing targets. It has no range capability, but gives the pilot the direction and speed of approach of a threat, or threats.

Range-while-search, used for detection out to about 80nm (150km) of all targets, regardless of velocity, aspect, or heading.

Track-while-search, which keeps tabs on up to ten separate targets and displays the eight most likely ones, while displaying the altitude, aspect, and velocity of the greatest threat; range is below 40nm (75km). Coupled with Amraam, TWS mode will allow simultaneous multiple target engagement.

Single target track is automatically cued if the RWS mode is engaged and the target comes within range. It displays steering commands and data for weapon launch on the HUD while target aspect, velocity and altitude are shown on a cockpit multi-function display. This mode can be used to give SARH guidance for Sparrow.

Raid assessment uses Doppler beam sharpening to separate the elements of a tightly bunched formation. It can be used at ranges up to 30nm (55km).

As described, these modes are all pulse, for look-up and look-level; for look-down, pulse-Doppler is needed. A pulse radar mode looking down picks up all the clutter from the surface of the ground; over the sea it might have some look-down capability, but not a lot. Pulse-Doppler

Below: Air-to-air search and track modes available with the F/A-18's APG-65 radar, showing relative ranges and scan angles in pulse look-up/look-level operation.

APG-65 air-to-air modes

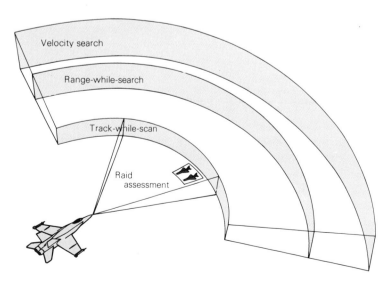

APG-65 displays

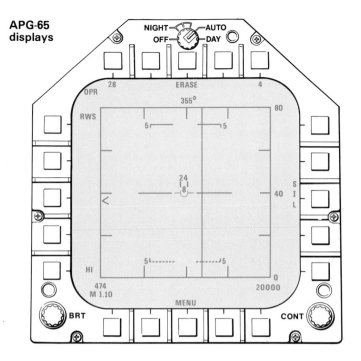

Above: Typical display generated by an APG-65 radar in range-while-search mode at high PRF and at a range of 80nm (150km).

Below: In track-while-scan mode the APG-65 can monitor up to ten targets while displaying data on the most immediate threat.

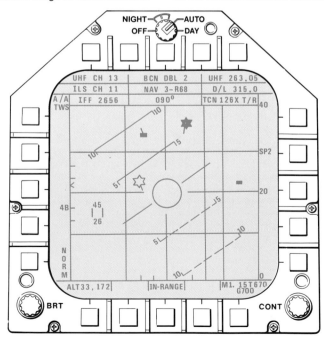

APG-65 boresight mode

Boresight acquisition uses a 3.3° beam aligned with the Hornet's centreline. This mode is most useful in a stern pursuit attack.

APG-65 vertical acquisition

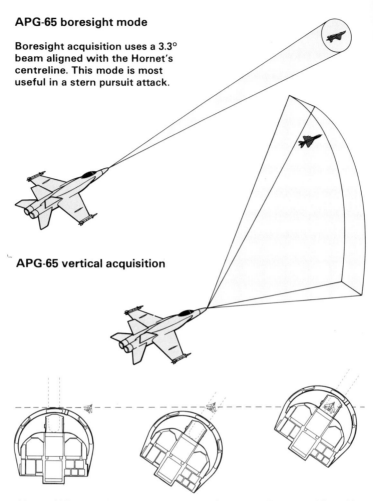

Above: With the Hornet or both it and the target in a hard turn vertical acquisition mode would be employed. While the radar scans an arc 5.3° wide 60° above and 14° below boresight, the F/A-18 pilot rolls his aircraft into the same plane of motion as the target — that is, he positions his aircraft so that the target appears to be above the centre of his canopy bow. Acquisition should then be automatic, but if necessary he can tighten the turn, causing the target to appear to move down.

screens out the unwanted radar returns and gives a good look-down capability, albeit at a certain loss of range capability. What it also does, and this is undesirable, is to screen out targets that are moving at about 90° angle off the fighter's flight path.

It should also be made clear that radar look-down is one thing, while missile shoot-down is another: to detect a target is not necessarily to be able to shoot it. The missile seeker head needs special filters to screen out the unwanted returns, and this is easier said than done, although the problem has been partly solved.

In addition to those already described, there are a series of air combat manoeuvre modes. These are **boresight,** which is a narrow beam pointing ahead of the axis of the fighter; **vertical acquisition,** which is used against a hard turning target; and **HUD acquisition,** which scans

APG-65 HUD acquisition

Head-up display acquisition mode scans the 20° by 20° HUD field of view once every two seconds, automatically locking on to the first target detected.

APG-65 gun director

Gun director mode uses pulse-to-pulse frequency agility to track targets and display a HUD gun aiming point: the pilot manoeuvres to place this on the target.

the area in the field of view of the HUD. All three are primarily for use with Sidewinder, although other missiles can be employed. Finally there is **gun director** mode, which is self explanatory.

The fighter radar traditionally scans a pie-shaped piece of sky ahead of it approximately 65° to each side of the centreline, although in some modes the angle is much smaller. The depth of the pie is quite thin; scan depth is divided into bars, and varies from two to eight bars. As a rough approximation, a bar is 2° deep, so the scan varies between 4° and 16°. The larger the field of scan selected, the longer the radar takes to sweep through it, the time variation being of the order of between a quarter of a second to 12 seconds or

more. The scan can also be trained up, level, or down.

The point is that the fighter radar can only cover a small portion of the sky in front of it, and none of the sky around, above and under it. In this respect, it is very limited; look-down allows a higher assailant to slip in unobserved, and vice versa, while the flanks and rear are unguarded. This is where AEW&C and JTIDS can prove decisive, keeping a friendly eye on a fighter's unguarded sectors.

Fortunately, the opposing fighters will also be heavily reliant on radar for early detection, and radar warning systems (RWS) are used to detect their emissions and give the pilots warning of what is going on, with azimuth and possibly altitude information, but probably no range. In a

27

confused multi-bogey situation this can be a mixed blessing, and the RWS needs to be programmed to give warning only of imminent threats.

Infra-red can be used for detection and has been for many years. It has the advantage that it is a passive detector, with no tell-tale emissions, but it is far from satisfactory in cloud-laden skies, and while it has better angular resolution than radar, it has proved unsatisfactory in use in the West, although it appears that the latest generation of Soviet fighters carry IR detection systems. Either the Soviets are more advanced in this field, or they are concerned about the latest American stealth projects: if the countermeasures on the B-1B are half what they are cracked up to be, IR detection might be better than nothing.

Finally, there remains visual detection, which in close combat will be more used than any other system. This has led to an all-round improvement in visibility from the cockpit, with teardrop canopies giving a good view to the rear. Mirrors on the canopy bow are also standard. Visual detection is essential at very short range; some smaller fighters are all

Below: An unseen opponent is detected and its position indicated on the head-up display of an F-14 Tomcat.

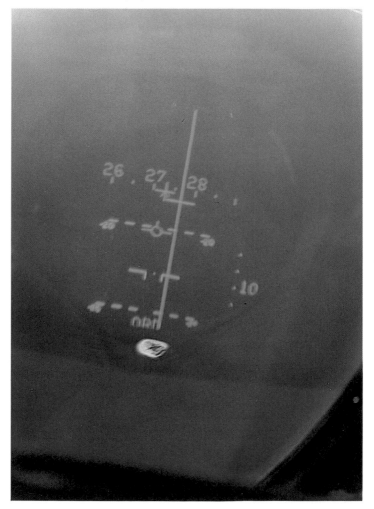

but invisible from head on at 2nm (4km).

We have examined the weapons and the detection systems, and they look pretty formidable. The case is the same for all opponents, as even if Soviet material and technology is behind that of the West, they will be level pegging within a few years, and the hardware flying now will be opposed by equal quality at some future date. The importance of the weapons and detection systems is underlined by the fact that many nations are refurbishing what can only be described as elderly airframes with new avionics to do battle in up to 15 years time, when some of the designs will be approaching their half centenary.

The other justification for this practice is that escalating costs have reduced the affordable number of modern fighters, and a refurbishment job is a cheap and easy way of keeping up the numbers, which must not be allowed to fall below a certain strength no matter how capable their replacements. There is at least one case of a Phantom pilot flying the same machine as his father flew many years before.

The modern fighter has to defeat weapons and systems the equal of its

Below: Another Tomcat is caught in an F-14 HUD. The extreme angle of bank is shown by the long oblique line.

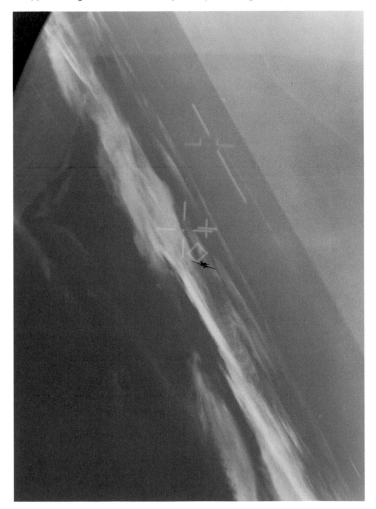

own, or sometimes inferior systems and aircraft but in superior numbers. What sort of machine is needed to do this?

The worst case situation in fighter terms is always an all-out war in Central Europe using conventional forces. Nuclear war has not seemed likely for many years, and since the Chernobyl incident seems even less likely. On the other hand, it seems equally certain that armed clashes will occur on a limited scale in other trouble-spots around the world, with the Middle East as hot favourite.

Fighter forces in both Western and Eastern Europe can be seen as a deterrent, helping to keep the admittedly uneasy peace, rather than an outright weapon: both the interceptor and the counter-air fighter are essentially defensive, in that their purpose is to neutralise attacking air units. They do, however, have to be capable of doing the job, and if successful, they must carry the fight to the enemy.

Taking the worst case first, conventional war in Central Euope, the Achilles' heel of almost all modern fighters is their dependence on a fixed base with long concrete runways. Such bases are vulnerable in the extreme, and it is debatable how many aircraft would leave the ground by the third morning. Good short field performance is therefore a prime asset, and the ability to operate from dispersed bases is even better.

The intake covers on the Soviet MiG-29 appear to be designed as anti-FOD shutters, which will enable this aircraft to take off from really rough strips, or from damaged airfields. The MiG-29 needs a braking parachute on landing, however, so the problem of recovery on a damaged strip remains. Alone among combat aircraft now flying, the Harrier will have little or no problem in this direction, and the Advanced STOVL project would appear to make more sense than the Eurofighter, the Rafale, or for that matter the American Advanced Tactical Fighter, although there is a possibility that the last programme might include a STOVL proposal.

Assuming that the fighters can get off the ground, much will depend on the tactical situation and how well the communications are working, as they will almost certainly be subjected to heavy jamming.

The overall tactical situation will be either defensive or offensive. In the defensive case, circumstances favour the defenders. They will be operating over their own lines, where enemy jamming of communications and radar will be less effective; the air

Right: Artist's impression of a possible ASTOVL fighter using vectored thrust to negotiate a bomb-cratered runway.

Below: Half a billion dollars' worth of 1st TFW F-15s parked on acres of vulnerable concrete at Langley Air Force Base, Virginia.

defence ground environment should be more or less intact; and the fighters will receive a fair amount of aid both from AEW&C aircraft and from ground control (one of their primary tasks will be to defend these vital facilities).

In the offensive case the probability is that the fighters will have to be able to work autonomously, depending entirely on their own on-board avionics and tactical skill. Each circumstance will have to be met as it arises; there will be little opportunity for preplanning. Their task will be to wrest a degree of air superiority over a limited area long enough to allow the interdiction units to do their work. Probably the only factor which will work for them is confusion, and there will be plenty of that. A confused situation is where superior training and tactical skills pay handsome dividends.

The first condition to be met in combat is to detect first, preferably avoiding detection at the same time, and in the defensive case, this is more easily achieved. If AEW&C and ground radars are working, the fighter pilot can put his radar on standby and let the defensive system vector him into an advantageous position, using his own radar only when an attacking position is about to be reached. At the same time, a sharp visual lookout is advisable in case an intruder has slipped the net.

Over hostile territory, these advantages no longer apply, and it is probably best to use the radar in intermittent scans to keep a watchful eye on events before latching on to a target. As we have seen, the range capability of radar is degraded when looking down, but at the same time, the fighter can travel faster at altitude. If mission requirements allow, it is preferable to remain at very low level over enemy territory with the radar looking up for optimum range performance, at the same time degrading the enemy's range performance. If detected, it can only be by pulse-Doppler radar, and the RWS should indicate this. A smart turn of 90° angle off to the enemy aircraft's flight path should then remove one's blip from his radar screen.

Over friendly territory, medium or high altitude may well be better, as it

APG-66 detection ranges

The detection range of the F-16A's APG-66 radar is substantially reduced in look-down mode. Figures for Soviet aircraft are calculated from flight test results involving US types.

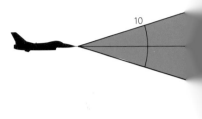

Below: Low and fast is safest in hostile airspace. The place to be is demonstrated by a pair of Eagles in the Grand Canyon.

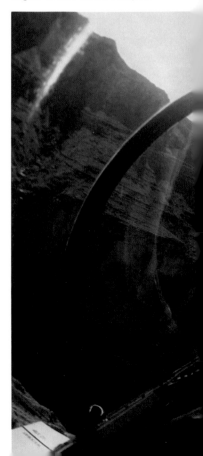

Range (nm)

80

70

Tu-95

MiG-25

MiG-23

F-4

30

MiG-23

F-111

MiG-25

Tu-95

Look-up

40

50

60

allows higher speeds and lower fuel consumption. If positioning for an attack using information from AEW or GCI is possible, the position can be reached much faster, especially if shoot-down missiles are carried. A front quarter attack is optimum if using SARH weapons, while the stern quarter is the preferred angle for heat seekers, even though they may have some all-aspect capability. If no shoot-down missiles are available, it becomes preferable to attain a level slightly below that of the target in order to cut out clutter. The main drawback of going to low level is that of missile range being reduced, making it necessary to get closer to achieve optimum firing range.

The easiest way to defeat an enemy missile attack is to stay out of his missile parameters: what is not launched can not hit. But in any shooting war it is almost inevitable that an air-to-air missile will be launched against a fighter. Once it is on its way, it cannot be stopped, but it can be defeated. The trick is to know that it is coming, and from what direction. It also helps to have some idea what type of missile it is and what its capabilities are.

A good rule of thumb is that a missile must be able to pull at least five times the g loading of a manoeuvring target to have a really good chance of success. A hard turn collapses missile envelopes quite dramatically, and if started before launch, may take the target aircraft out of parameters before the launch can be accomplished. If not, it can still make tracking very difficult for the missile.

Countermeasures are obligatory, chaff against a radar homer and flares against a heat seeker, or both if the defender is not certain. They may work, but they may not, and no true fighter pilot is going to stake his life on countermeasures when he can take further action.

If it is thought that the missile is a heat seeker, it is a perfectly valid move to fly into a cloud if there is one handy and it can be reached in time, but if the possibility exists of the adversary aircraft carrying radar homers also, it would be wise to get out of the cloud and regain sight as soon as the missile has been shaken

off. Clouds apart, the best move is to turn the hot tailplane away from the missile, and reduce the plume of hot exhaust gas by throttling back. Some missiles are designed to home on the hot metal of the tailpipe, while others

Right: Overwing launch of a Matra R550 Magic heat-seeker by a French Air Force Jaguar.

Below right: IR image of flares fired by a USAF Phantom shows the multiple targets produced.

Below: An R550 Magic demonstrates its ability to deal with a manoeuvring target.

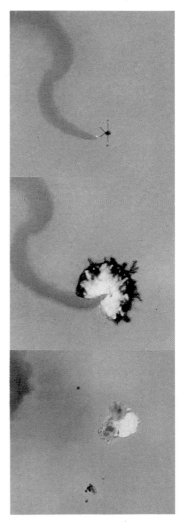

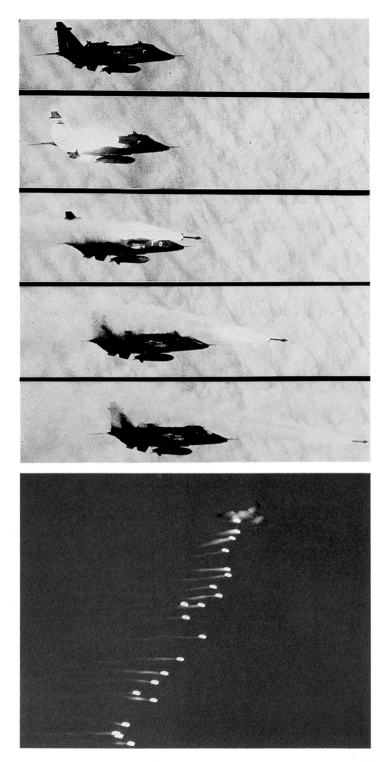

35

use the exhaust plume as a target. Aircraft which use reverse thrust are more vulnerable to the former, as it is impossible to shroud the tailpipe as is done on some aircraft. Finally, lining the aircraft up with the sun can sometimes decoy a heat missile.

Against a radar homing missile, a 90° angle off may work if it is a shoot-down type, while if all else fails, maintaining a beam-on aspect will take advantage of any fuzing problems it may have, as previously noted.

Manoeuvring makes any type of missile work harder to track. Most missiles use proportional navigation in their homing systems, which operate by manoeuvring the missile to fly a constant lead path on a collision course. A hard turn gives this far more work to do and, better still, may cause the movable seeker head to bump its gimbal limits and lose track altogether. The object of the exercise is to give the missile seeker head a rate of angular change of the line of sight which is too high for it to cope with.

Sometimes the missile is not seen until too late to develop enough angle off, but a hard break downward using gravity for extra assistance may rectify matters. This has a secondary advantage; it forces the missile seeker head to look down into the ground clutter, and it may break lock. If the missile continues to track the fighter downward, a last ditch hard break back up and out of plane can often throw it off.

A rear quarter shot from near maximum range can often be outrun, although this sounds distinctly sporty, and it is necessary to keep the missile in sight at all times, so a small degree of angle off should be maintained. On the other hand, a near minimum range front quarter shot can be defeated by a turn into the missile and across its nose. There is always a slight lag in guidance system response times, and as it turns to follow, a hard reverse by the target fighter is likely to cause an out of phase response as the guidance system tries to correct, resulting in a wide miss. This is most effective at high altitudes where the control response times of the missile are at their slowest.

Countermeasures apart, certain factors can also be built into the fighter at the design stage. These are the so-called stealth features which reduce the detection signatures, and with them the range at which the fighter can be detected. The radar signature can be reduced by avoiding sharp corners and angles of 90° or less; wing-body blending is a good stealth feature, as is shielding the face of the engine compressor if this can be done without compromising engine performance. The use of radar absorbent materials and composites also help.

The infra-red signature is rather more difficult to hide, although chemicals can be used to lower the temperature of the exhaust gases. The Sea Harrier, with its four effluxes, has a larger exhaust plume than

Right: Replay of an engagement on the Cubic Combat Manoeuvring System. A symbolic coffin encloses the victim of an AIM-7F.

Below: Cockpit and display in BAe's air combat simulator, with a Phantom in the process of engaging a MiG-23.

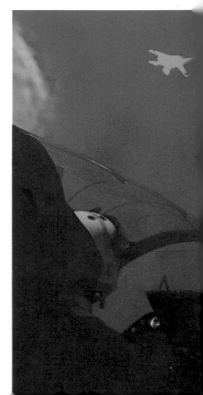

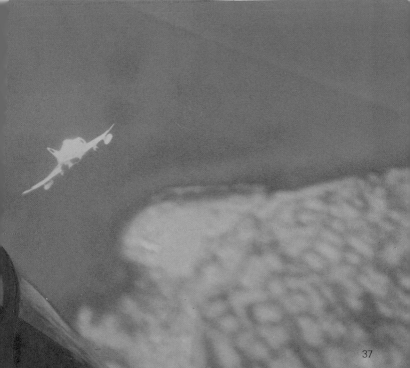

most, but it is more diffuse, and in many ways forms a more difficult target.

Probably the most difficult of all to hide is the visual signature. The adoption of low visibility paint is fine, but it only caters for a certain range of backgrounds. White paint in the intakes reduces the black hole effect of a fighter approaching nose-on, but the best answer is to make the fighter small, even at the expense of some capability.

As mentioned earlier, the best way to defeat a missile — or gun — attack is to keep out of firing parameters. And however superior the fighter's detection and weapon systems may be, it will become necessary sooner or later to outfly the opponent.

Often there will be no alternative to encountering an enemy fighter from head-on. It is a valid ploy to launch a missile out of parameters under these circumstances, for the sake of getting the first shot in: it would be a foolhardy opponent that would stake his life on the shot not being a threat, and his normal reaction will be to forget offensive action and concentrate on defeating the oncoming missile. He will be forced to manoeuvre on the defensive, and with his attention concentrated on the oncoming missile will probably lose track of the fighter that launched it. This will give the latter the initiative instantly, enabling it to start manoeuvring into a position of advantage.

Manoeuvre combat is conducted at visual distance, although sometimes this will be marginal. Normally it will be multi-bogey, with several fighters milling around and a distinct possibility that others, from both sides, may join the fray from any direction at any moment. The well trained fighter pilot has a whole armoury of air combat manoeuvres at

Above right: Many air forces have experimented with attitude deception schemes, but only the Canadian Armed Forces have adopted false canopies.

Right: A more subtle scheme was this asymmetric two-tone grey camouflage applied to F-14s for trials in 1977.

his disposal, but rarely will there be time to employ more than one or two before the need to clear his tail arises.

It is commonly thought that the way to handle a multi-bogey situation is to hang around on the edge of what is termed the bogey cloud, and try and pick off anyone who gets isolated from maximum range, but the dynamics of modern air combat mean that the fighter is sooner or later going to get caught up in it.

No matter how good the pilot, and how good his aircraft and avionics, he is inevitably going to lose track of the broad situation. He will not be alone, although wingmen may well be forced to split off, and provided there are several friendly fighters on the scene, a certain amount of coincidental support will be provided, so that an opponent turning in behind may well find that he in turn is coming under attack.

The main prerequisite of launching the attack is to obtain surprise, and the second to select the egress route. The attack is made at high speed at a preselected target, after which the profile is a series of short hard turns of between 60° and 90°, interspersed with brief straight line accelerations to keep energy levels high, and taking shots of opportunity on the way. To take time out to tackle as a single opponent for more than a few seconds would be to risk retribution from his friends, and get caught up in the 'furball', from which, considering the capability of modern missiles, it will be difficult to extricate oneself. The dogfight in the World War I sense no

Right: Even in dry thrust the F-16 has impressive performance. In combat, the use of afterburner offers brute force manoeuvre solutions while an opponent might need to conserve energy.

Below: With afterburners lit and wing flaps deployed a Tornado ADV begins its climb to operating height. Good climb rate is essential to minimise the time required for an interception.

41

longer exists, and any attempt to emulate it might well prove fatal.

We have covered the weapons and the detection systems, and the main circumstances in which they are likely to be used. What then are the qualities needed by the modern fighter, and by this, we mean a fighter that will remain a formidable adversary in say, 15 years time?

The F-4 Phantoms of the Luftwaffe and those of many other nations are about to be upgraded, yet their aerodynamic capabilities remain those of the 1960s. In the multi-bogey situation that we have just outlined, they remain a very potent weapon, using 'blow-through and shoot' tactics, but if they do get split off and caught in the furball their chances of survival are not good: they simply do not have the performance and manoeuvrability to fight their way out. The same applies in limited wars where the numbers taking part are small. The smaller the force density in any engagement, the better the aerodynamic and handling qualities of the fighter need to be.

Most future air fighting will take place at low and medium altitudes, so a high ceiling is only of value to intercept the occasional high flying reconnaissance aircraft or intruder. A good climb rate is essential to enable the fighter to get off the ground and reach operating altitude as rapidly as possible.

Climb rate can be approximately linked with acceleration, another must. Acceleration gains energy quickly for the attacking fighter and also for its missile on launch, giving the missile a better chance of scoring. It helps to confer the initiative, and also facilitates escape from an unpromising situation. Finally it replaces energy which has been bled off in hard manoeuvring.

Mach 2 is useful for an interceptor but not for much else. It takes forever to get there, and uses a lot of fuel when it does. In combat there would rarely if ever be time to accelerate from subsonic cruising speed to Mach 2. On the other hand, a supersonic top speed at sea level with a full bag of weapons would be well worth having.

Much has been written about turning ability, and especially sustained turn, in recent years. The fact is that the turn rate of modern fighters is pretty slow by World War I and II standards, while the turn radius is vast. This is because turning ability is closely linked to speed, and speeds have increased manyfold. In a combat situation, the best turning fighter is often the one that is the slowest at that moment, limited only by corner velocity, which can be defined as the point where lift limitations and structural limitations meet; it is therefore the point where the highest rate of turn, and normally the lowest radius of turn, are achieved. Above corner velocity, the minimum radius increases rapidly but the turn rate decreases slowly. In combat it is therefore better to start at a speed well above corner V, as speed will bleed off rapidly down to this level.

Sustained turn is exactly what it says, a turn rate and radius that can be maintained for long periods. Naturally it varies with altitude. Instantaneous turn is the turn rate and radius that a fighter can pull in exactly one set of conditions, and it will change as the speed winds down. It is

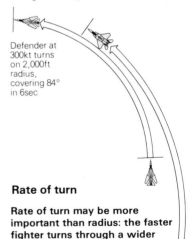

Defender at 300kt turns on 2,000ft radius, covering 84° in 6sec

Rate of turn

Rate of turn may be more important than radius: the faster fighter turns through a wider radius, but turns 90° against the target's 84°.

Attacker at 500kt turns on 3,200ft radius, but covers 90° in 6 sec

Turn radius and turn rate

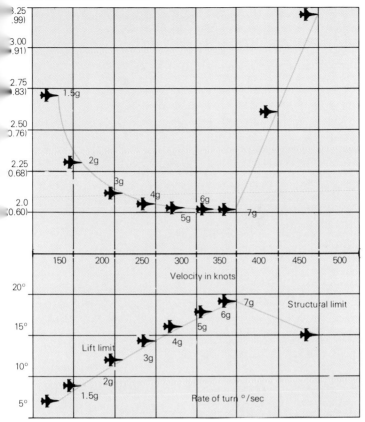

Above: For a fighter with a minimum flying speed of 150kt and a strength limit of 7g both radius and rate of turn are best at around 400kt.

Below: In a maximum power turn initiated at a speed of Mach 1.2 in formation with an F-4, an F-16 demonstrates its ability to turn on half the Phantom's radius.

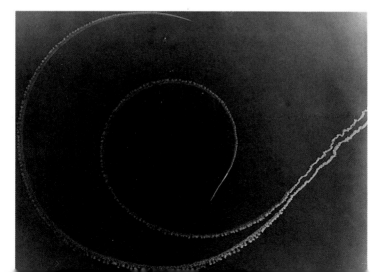

in most cases superior to sustained turn and will be more readily used in emergency. But what is more important than either is how quickly the fighter can get into the turn.

A fast rate of roll is good, but rate of acceleration into the roll is even more important, as it determines how quickly a turn can be initiated. Pitch rate is another factor here, and speeding this up is just one reason why the canard foreplane has become so popular in recent years. Dassault, who use it on Rafale, are convinced that instantaneous manoeuvre capability is more valuable to a fighter under modern combat conditions that sustained performance.

The next important point is a good thrust:weight ratio, preferably better than unity on takeoff. This improves climb rate, and confers a good sustained turning performance, and most important, aids manoeuvre generally through a high level of specific excess power under a wide range of flight conditions. This increases manoeuvrability; it also serves to replace energy bled off in a hard-turning fight very quickly.

Finally we have combat persistence. In part, this is the number of on-board kills carried; two missiles would not last long in a multi-bogey fight, and ideally a mix should be carried of radar and heat homers for operational flexibility. The optimum number seems to be emerging as about six or eight. On-board fuel is yet another factor in combat persistence. Fuel may be carried externally in jugs, but they hinder performance, and often have to be dropped in action. They also sterilise a pylon which might otehwise have been used for weaponry. The optimum fuel fraction seems to be emerging as 0.3 or a little under. More fuel needs

extra structure and volume to hold it, all of which adds weight. The weight of the fuel itself also affects the thrust/weight ratio.

The current state of the art fighters are both in the technology demonstrator stage. They are the Eurofighter, as evidenced by the British

Right: BAe's EAP demonstrator takes off for the first time on August 8, 1986. Described by test pilot Dave Eagles as "superb, remarkably agile and yet very easy to fly . . . ideal and what every fighter pilot would want," the EAP shares the unstable delta canard configuration with Dassault-Breguet's Rafale and points the way to the future European Fighter Aircraft.

Aerospace EAP, and Dassault's Rafale. Each is an unstable canard configured delta with a kinked leading edge to the wing, a moderate wing loading and low aspect ratio, a high thrust loading and all the latest electronics. Both are optimised for very high manoeuvrability, both sus-tained and transient. Similar in some ways, in others they show markedly different solutions to what is essen-tially the same problem, perhaps the most obvious of which is the design of intake. Where fighters of the near future are discussed, these are two to watch.

Fighter Aircraft

It is customary in books of this nature to list current fighter aircraft together with a greater or lesser amount of data culled from manufacturers' brochures or other sources. Some of the data is accurate; some is accurate but not particularly relevant; yet more is of a "gee whiz" nature: the difficulty for the reader is in making valid comparisons between the different fighters shown.

There are many reasons why this is so. Often the information, while accurate, lacks relevance. A fighter may well be able to attain a speed of Mach 2.5 in a clean condition, but without missiles it would be unable to harm an opponent. It could use its speed to escape from an opponent after exhausting its missiles, but would take two or three minutes to wind up to top speed anyway, during which time it would be vulnerable. Top speed, or V max, is irrelevant unless it is attainable by a fighter carrying suitable air to air weapons, in a tactical situation that permits the relatively long acceleration period needed. Most other brochure parameters are also irrelevant, for much the same sort of reasoning.

In other areas the data in the brochure is accurate — or is it? Take wing area: for American aircraft the wing area usually includes the body area between the wings, but this is not the case for aircraft of other nations, so wing area and the closely allied aspect ratio differ as a result. Even length is subject to variation: taking a tape measure to an aircraft and measuring it physically, from front to rear, would probably produce figures substantially different from those in the manufacturer's brochure not because of any lack of accuracy on either part, but as a result of some house rule which dictates how the measurement is to be conducted. The British Aerospace Lightning is a case in point; the house rule states that the length shall be measured from the lip of the intake rather than from the tip of the projecting centrebody.

Other discrepancies arise from a lack of accuracy in compiling the brochure figures. Often Imperial and metric measurements do not tally, in which case the figures given in the system used in the country of origin are most likely to be right. There are also, regrettably, instances of terminological inexactitudes, the purpose of which appears to be to make the product look superior to its rivals. Manufacturers are, after all, in the commercial marketplace, where selling the product is the name of the game, and it would be unrealistic not to expect a bit of sales hype to creep in somewhere.

Another major factor in inaccuracy is security. The product has to be sold, but the customer is not going to be best pleased if a potential adversary, who is equipped with a rival product, has been able to discover exact details of what the new machine will do and, more importantly, what it will not do. Brochure material is therefore of necessity rather bland; it must give little away.

Fashion also plays its part, even with the engines of war. Thirty years ago, a V max of Mach 2 was considered absolutely essential, and guns were passé. Ten years ago sustained turn rates, guns and 9g loadings without pulling the wings off were the fashion. Today the essential attributes are agility, carefree manoeuvring and pointability.

The final point about both brochure and press release material is that it is almost impossible to make it simultaneously accurate, relevant and understandable by the non-expert. Flight is a dynamic process, and capabilities of an aircraft are changing constantly from takeoff to touchdown. Factors that affect performance include speed, altitude, ambient air temperature and pressure, while the weight of the aircraft is gradually reducing as fuel is burned off and missiles launched. Drag varies by a tremendous amount, not only with speed, but also with the angle of attack in manoeuvring flight and the external weaponry and loads carried. Finally, data such as the internal fuel capacity is often not given, and this is most important for making valid comparisons between different types.

Yet another glaring omission is the drag coefficient. In a world where drag coefficients are the latest selling points for the motoring fraternity, this seems difficult to understand, but maybe it would be giving too much away. It is also reasonable to assume that too much data would be counter-productive.

Given such limitations, we have endeavoured to formulate a simplistic approach to the evaluation of fighter aircraft from the non-classified information which is freely available but without becoming immersed in advanced aerodynamics. In certain cases, notably for the newer products originating east of the Brandenburg Gate, it has been necessary to attempt what can best be described as a guess tempered with commonsense, although this has been kept to a bare minimum. Certain brochure figures have been

omitted from the data tables, although where these are informative, they have been included later in the text.

The format for the tabular data is outlined here, and the reasons for its adoption given where necessary. All data is given in both Imperial and metric measures, in that order. Fighters are listed in chronological order by date of prototype first flight, as another aid to making true comparisons, although where certain types have run into more than one generation this has been noted.

Dimensions
Length, wingspan and height are given in feet/metres. Wing area is given in square feet/square metres. Aspect ratio is the ratio of the square of the wingspan to the wing area.

Weights
These are given in lb/kg, are approximate and are rounded off. Empty weight is generally the brochure figure where available. Take-off weight is in fighter configuration with guns loaded, maximum internal but no external fuel, and the normal load of air-to-air weapons, as stated in the text. Combat weight is as take-off weight but with only 50 per cent inter-

Below: By any standards the F-15 is one of the most formidable fighters in existence, though simple tables of figures do not tell the whole story.

nal fuel, is generally calculated from brochure data and is approximate. Gross take-off weight has not been included: it is more relevant to the ground attack mission than to the fighter role, and in any case it is usually a theoretical figure totalled from the basic aircraft weight with the rated loads of the weapons hard points added. It can rarely be achieved in practice.

Power
Engine make and type, with thrust outputs at maximum (max) and military (mil) settings. The outputs are given in terms of static thrust at sea level in lb and kN. Of course, in the dynamic conditions of flight these alter considerably, often being greater at low altitudes but inevitably dying away as height increases. It should be noted that turbofans are much more economical at cruise settings than the older turbojets.

Fuel
Given in lb/kg. Where only volume figures have been available, the weight of fuel has been calculated as JP-4 at 6.5lb/US gallon, for the purposes of standardisation. Internal fuel is always given, external sometimes only where it seems relevant to do so, such as where long patrol times over friendly or neutral territory are required. When making comparisons, external fuel should be considered with caution as it often sterilizes a pylon which could otherwise have carried an AAM, and is therefore adding to flight endurance but reducing combat endurance. By the same token, in-flight refuelling is at times a tremendous advantage, but is impractical within range of enemy fighters. Finally, the fuel fraction is the percentage of fuel expressed as a proportion of the take-off weight in fighter configuration. Figures of 0.28-0.30 indicate a fighter has acceptably long legs; a lower figure means it is more or less short on operational radius, while a higher figure means it is suffering a considerable weight penalty, not only in terms of extra fuel but also for the weight of tankage and added structural strength necessary to carry it. A turbofan-powered fighter will achieve a better radius performance

Above: Accurate data on modern Soviet types is hard to come by, and the first appearance of the MiG-29 in the West produced numerous conflicting evaluations.

for the same fuel fraction than one powered by a turbojet, while an unaugmented turbofan-powered fighter will do best of all.

Loadings
Loadings are divided into thrust and wing. Thrust loadings are expressed as a proportion of the engine static thrust against the weight; it has become fashionable during the last decade for fighters to have thrust loadings in excess of unity, or l:l, which usually denotes a high level of SEP (specific excess power) with which to manoeuvre or climb or accelerate. In practice it is not quite that simple, as drag is also an important factor, but it does give some indication. Tornado F.3, to give but one example, has quite a sorry thrust: weight ratio, but its exceptionally low drag coefficient gives it surprisingly good performance — better, in fact, than that of certain fighters with a considerably better power loading.

It is usual to give power loadings at combat weight, using that as an optimum point, but for comparative purposes that supposes that all fighters are down to 50 per cent internal fuel when they engage, which is patently absurd. It is conceivable that a fighter in the air superiority role could be engaged almost as its wheels left the concrete, in which

case it would probably jettison any external fuel tanks and wade in. On the other hand, a fighter reduced to 50 per cent internal fuel is in many circumstances going to fight with one eye on the fuel gauge. Such is the spread we have chosen in our tabular material, from take-off to combat weight.

We have also chosen to give the loadings in mil power as well as max, and examination of the tables will show that the power loadings of some fighters in mil exceed those of others in max. To take a specific case, a laden Flogger would hardly force an F-15 to use full 'burner. It is also possible that fuel shortage may be the critical factor in how the fight is conducted.

Wing loading is stated in lb/sq ft (kg/m²), and the same weight range has been selected, from take-off to combat weight. Wing loading has traditionally been a method of assessing turning capability, but it is not quite so straightforward: aspect ratio, wing shape and high-lift devices also play a part; lift coefficient is another factor, but not a large one if supersonic speed is required. A low wing loading is just one of three factors that go to make up good turning capability, the others being a fairly high aspect ratio and a high thrust loading.

Performance
V max is stated in terms of Mach number, and is given both for high altitude and for low level flight. It should be remembered that Mach 1 is

88kt (163km/hr) greater at 36,000ft (10,973m) than it is at sea level.

Service or operational ceiling is stated in ft/m. Initial climb rate is given in ft/min (m/sec), and is attained only at about Mach 0.9 at sea level. Take-off and landing roll are given in ft/m. All other performance data, such as time to altitude, acceleration, instant and sustained turn rates and combat radius, are given in the text where they can be suitably qualified.

It should be remembered that V max and ceiling are irrelevant to combat, and that they are included mainly as a reminder of where the top right hand corner of the performance envelope lies: the true figures are at least 20 per cent less depending on the missile load carried. No recorded combat has involved speeds in excess of Mach 1.6, and very few have taken place above 30,000ft (9,144m) since the Korean War more than 30 years ago.

First Flight
The date given is generally that of the model to which the data applies, though the aircraft are arranged in chronological order according to the date of the prototype's first flight.

General
It should be noted that the most reliable available figures have been quoted, though in some cases they have been modifed and in others they have been calculated by the author using known details.

Mikoyan MiG-21 Fishbed

Type: Single-seat point defence fighter with limited adverse weather capability (some versions), limited ground attack capability. Reconnaissance and two-seat trainer version exists.

The MiG-21 was conceived in 1953, at a time when the greatest threat was the nuclear-armed manned bomber: the American B-47 was entering service in large numbers, and with the B-52 already on the horizon the primary requirement was something able to intercept them. But without unlimited resources, this was a conflicting requirement. A

fighter was needed that could take off quickly, gain altitude with a minimum of delay, and make the interception. A high rate of climb, a service ceiling of 65,500ft (20,000m) and a speed of Mach 2 at altitude were called for, together with reasonable manoeuvrability and handling.

So far, so good, but it also had to be available in large numbers, since the potential threat force was numerically strong and the Soviet Union is a large area, with a correspondingly large perimeter to defend: that meant it had to be cheap and easy to manufacture. At a time when much of the West was thinking in terms of big, colossally expensive

Dimensions	MiG-21F Fishbed-C	MiG-21MF Fishbed-J	MiG-21bis Fishbed-N
Length (ft/m)	51.71/15.75	51.71/15.76	51.71/15.76
Span (ft/m)	23.46/ 7.15	23.46/ 7.15	23.46/ 7.15
Height (ft/m)	13.45/ 4.10	14.75/ 4.50	14.75/ 4.50
Wing area (sq ft/m²)	247/22.95	247/22.95	247/22.95
Aspect ratio	2.23	2.23	23
Weights			
Empty (lb/kg)	10,800/4,900	11,850/5,370	13,500/6,120
Takeoff (lb/kg)	16,250/7,370	18,080/8,200	19,730/8,950
Combat (lb/kg)	14,140/6,410	15,580/7,070	17,230/7,820
Power	R-11	R-13	R-25
Max (lb st/kN)	13,670/60.8	14,450/64.2	19,840 0/88.2
Mil (lb st/kN)	9,600/42.7	11,240/50.0	12,790/61.3
Fuel			
Internal (lb/kg)	4,220/1,910	5,000/2,260	5,000/2,260
External (lb/kg)	840/380	840/380	840/380
Fraction	0.26	0.28	0.25
Loadings			
Max thrust	0.84 0.97	0.80 0.93	1.01 1.15
Mil thrust	0.59 0.68	0.62 0.72	0.65 0.74
Wing takeoff (lb/sq ft/kg/m²)	66/321	73/357	80/390
Wing combat (lb/sq ft/kg/m²)	57/279	63/308	70/341
Performance			
Vmax hi	M = 2.00	M = 2.05	M = 2.10
Vmax lo	M = 0.9	M = 0.98	M = 1.01
Ceiling (ft/m)	57,400/17,500	59,000/18,000	59,000/18,000
Initial climb (ft/min/m/sec)	25,900/132	N/A	55,775/283
Takeoff roll (ft/m)	N/A	N/A	N/A
Landing roll (ft/m)	N/A	N/A	N/A
First flight	Dec 1955	1967	1975

fighters, capable of Mach 3 and carrying enough firepower to sink a battleship, the Soviet Union opted for simplicity and numbers. In any case, Soviet technology was behind that of the West.

The Ye-4 tailed delta, which can be regarded as the first prototype of the MiG-21, first flew in December 1955. The delta wing planform had many advantages in high-speed, high-altitude flight, and was easy to manufacture. It also had no clearly defined point of stall, and developed maximum lift at very high angles of incidence, albeit at a high penalty in increased drag. Drag was also high in manoeuvring flight, but the tailed delta avoided the worst drawbacks of the tailless variety.

With the benefit of hindsight, the MiG-21 family can be regarded as a success. Its service life looks set to exceed 40 years, and it has been built in greater quantities than any other supersonic fighter, in at least as many subtypes, and it has been operated by more nations. It is also accorded a fair measure of respect in the fighter community. Yet its combat record is hardly inspiring, considering that it has been involved in almost every shooting war around the globe since 1965.

The truth is that it is a very ordinary fighter, and had it been of Western origin would probably have sunk without trace prior to 1970. As a pilot's aeroplane, it is regarded as simple to fly, and possibly the least demanding of all the Mach 2-capable types. The controls are heavy, to a ▶

Above: MiG-21bis Fishbed-L of the Soviet Air Force. Lack of range and poor visibility from the cockpit are two drawbacks.

Below: The MiG-21 has been widely used in the Middle East. This Egyptian aircraft strafes a range target with cannon.

▶ degree where a fair amount of muscle is needed, though this does help prevent the pilot from overstressing the airframe. The avionics fit is basic, with Spin Scan search and track radar in the early versions, and the more advanced but still short-range Jay Bird in later types, so it is dependent on good GCI to be really effective. The pilot's view out is not good, rear vision being almost non-existent, and even the view ahead is restricted by both avionics displays and a heavy canopy bow.

Perhaps the type's worst shortcomings arise from a design fault. With a fairly low fuel fraction it is a short-legged fighter, but the internal fuel is carried mainly ahead of the centre of gravity. As this burns off, the centre of gravity moves aft until it passes the limits of controllability, with the result that up to a fifth of the internal fuel load is unusable. This imbalance not only reduces the combat radius without external fuel to a ridiculously short distance — (the

MiG-21F on internal fuel had an interception radius of just 120nm (222km) — but in practice means that Mach 2 can not be achieved without running out of fuel, Mach 1.9 being about the limit, even in clean condition. It also means that in combat, the pilot is forced to fly with one eye on the fuel gauge. Nor can the brochure ceiling figure be achieved; a zoom climb will reach no more than about 46,000ft (14,000m), and performance above 20,000ft (6,100m) has been described as poor.

Turning ability has always been presented as the strong point of the MiG-21, and at the lower speed levels, it is pretty good by any standards. This stems partly from the low corner velocity, which is somewhere around 300kt (556km/h), and at Mach 0.5 and 15,000ft (4,600m) it can work up an instantaneous turn rate of 11.1°/sec, which is rather better than its frequent adversary, the F-4E Phantom, can manage.

On the other hand, increase the

speed to Mach 0.9 at the same altitude and the instantaneous turn rate has only increased to 13.4° and sustained turn is down to 7.5°/sec. Both of these are worse than the slatted Phantom's figures, though to be fair, they are better than the exactly contemporary Mirage III and Starfighter can achieve.

The latest Fishbed-N, with its far superior thrust:weight ratio, is likely to have a far better sustained turning performance. Manoeuvre flaps are used to aid turning in combat; they are extended to the takeoff setting at low speeds, and gradually bleed in under dynamic pressure from around 220kt (400km/hr) up to 380kt (700-km/hr) when they are fully retracted.

However, although the MiG-21 has not done very well in combat to date, it still must be regarded with respect. The MiG-21bis, with its more powerful engine, is a formidable dogfighter, especially if it is equipped with Western avionics. Whatever its drawbacks, it is popular, and it is popular because it is cheap. In terms of equal cost quantities, it is difficult to find better value.

Armament
Early versions: one 30mm NR-30 cannon and two AA-2 Atoll missiles. Later versions: one GSh-23 23mm cannon with 200 rounds per gun and two or four AA-8 Aphid missiles.

Users
Afghanistan, Algeria, Angola, Bangladesh, Bourkina Fasso, Bulgaria, Cuba, Czechoslovakia, Egypt, Ethiopia, Finland, East Germany, Hungary, India, Iraq, North Korea, Laos, Madagascar, Mongolia, Mozambique, Nigeria, Poland, Romania, Somalia, Sudan, Syria, Uganda, USSR, Vietnam, North Yemen, South Yemen, Yugoslavia, Zambia

Below: MiG-21bis Fishbed-N, with typical warload of two AA-2 and two AA-8 missiles.

Dassault-Breguet Mirage III/5/50/3NG

Type: Single-seat interceptor and air superiority fighter with limited adverse weather capability in some versions; others optimised for ground attack. Reconnaissance and two-seat trainer versions exist.

During the years following 1967 the name Mirage was synonymous with success. Outnumbered and surrounded, the Israeli Air Force had subdued the mainly Soviet-equipped Arab air forces during the Six-day War with devastating efficiency. It then maintained an ascendancy over its adverseries through the pro-

tracted War of Attrition, despite clashing with MiG-21s flown by Soviet pilots, until the short but vicious slogging match of the October War in 1973, when the Israelis were again victorious. Meanwhile, many nations clamoured to buy a combat proven fighter, especially one that, for a Western product, was affordable, easy to maintain, and not hedged around with restrictions like most American and British fighters.

The Mirage III was conceived at about the same time, and as a response to the same type of threat, as the MiG-21. The French aircraft in-

Dimensions	Mirage IIIC	Mirage IIIE	Mirage 5
Length (ft/m)	48.46/14.77	49.23/15.04	51.00/15.54
Span (ft/m)	27.00/ 8.23	27.00/ 8.23	27.00/ 8.23
Height (ft/m)	13.96/ 4.26	13.96/ 4.26	14.75/ 4.50
Wing area (sq ft/m²)	377/35.00	377/35.00	377/35.00
Aspect ratio	1.93	1.93	1.93
Weights			
Empty (lb/kg)	13,040/5,915	15,540/ 7,050	14,550/6,600
Takeoff (lb/kg)	19,000/8,620	22,050/10,000	21,300/9,660
Combat (lb/kg)	16,440/7,460	19,190/ 8,700	18,320/8,310
Power	Atar 9B	Atar 9C	Atar 9C
Max (lb st/kN)	13,320/59.2	13,670/60.75	13,670/60.75
Mil (lb st/kN)	9,460/42.0	9,430/41.9	9,430/41.9
Fuel			
Internal (lb/kg)	5,135/2,330	5,722/2,595	5,967/2,710
External (lb/kg)	5,860/2,660	5,860/2,660	5,860/2,660
Fraction	0.27	0.26	0.28
Loadings			
Max thrust	0.70 — 0.81	0.62 — 0.71	0.64 — 0.75
Mil thrust	0.50 — 0.58	0.43 — 0.49	0.44 — 0.51
Wing takeoff (lb/sq ft/kg/m²)	50/246	58/286	57/276
Wing combat (lb/sq ft/kg/m²)	44/213	51/249	49/237
Performance			
Vmax hi	M = 2.15	M = 2.2	M = 2.2
Vmax lo	M = 1.14	M = 1.14	M = 1.14
Ceiling (ft/m)	54,100/16,490	N/A	N/A
Initial climb (ft/min/m/sec)	16,400/83	N/A	N/A
Takeoff roll (ft/m)	5,250/1,600	N/A	3,500/1,070
Landing roll (ft/m)	N/A	N/A	N/A
First flight	Nov 1956	N/A	May 1967

dustry, virtually non-existent at the end of World War II, had clawed its way into the military fast jet market from scratch, determined to catch up with the UK and the USA.

With the fast, high-flying bomber the main threat, the Mirage III was developed as an interceptor. The delta wing planform was adopted for its high-speed, high-altitude qualities and for its ease of construction. It was intended to climb fast and intercept rapidly, and little if any thought was given to manoeuvre capability, the tailless delta being worse in this respect than the tailed delta form of the MiG-21.

The Mirage was fairly austere, and was an all-French solution to a French defence need. To give extra climb and extra speed at high altitudes, where the power of a turbo-jet is in decline, a supplementary rocket motor was incorporated, the fuel for which occupied the space for the gun ammunition, reducing the armament to a single, rather ineffective Matra R.511 missile carried on the centreline. This was later supplanted by the rather more effective R.530, and later still two R.550 Magics were hung beneath the wings. When the rocket was not fitted, the gun armament was two 30mm DEFA cannon with 125 rounds per gun.

Although designed as an interceptor, it was as a tactical fighter that the Mirage achieved fame. The view out was considerably better than the view from the MiG-21, though the rear view was almost as bad; the cockpit itself was austere but comfortable. The Mirage handled very nicely, and was definitely a pilot's ▶

Mirage 50	Mirage 3NG
50.42/15.37	51.33/15.65
27.00/ 8.23	27.00/ 8.23
14.75/ 4.50	14.75/ 4.50
377/35.00	377/35.00
1.93	1.93
15,800/ 7,170	15,540/ 7,050
22,650/10,275	22,050/10,000
19,670/ 8,920	19,060/ 8,650
Atar 9K50	Atar 9K50
15,870/70.5	15,870/70.5
11,060/49.2	11,060/49.2
5,967/2,710	5,967/2,710
5,860/2,660	8,060/3,660
0.26	0.27
0.70 − 0.81	0.72 − 0.83
0.49 − 0.56	0.50 − 0.58
60/294	58/286
52/255	51/247
M = 2.2	M = 2.2
M = 1.14	M = 1.14
60,000/18,300	54,000/16,450
36,600/186	N/A
2,600/800	N/A
N/A	N/A
Apr 1979	Dec 1983

Above: A Mirage IIIE of the Armée de l'Air displays the tailless delta configuration adopted for high-speed, high-altitude interception.

▶aeroplane, though it lacked stability at high altitudes; it could have used a bit more power and a better radar.

The Cyrano radar is generally considered inferior to the British Airpass, but an RAF exchange pilot with the Armée de l'Air feels that this was because the Cyrano was generally used over land, whereas Airpass was more often used over the sea. In all other departments performance was adequate; the standard tailless delta faults were that speed bled off alarmingly in hard turns, and take-off and landing runs were longer than they need have been.

After the Six-Day War, the Israelis proclaimed the Mirage to be the gun-fighting aircraft supreme, although given its inferiority in turning ability compared with the MiG-21 and attendant speed loss in hard turns, this can hardly have been the case. Infact, the Israelis found both the radar and the R.530 missile to be unreliable; the only R.530 kill recorded occurred on November 29, 1966, when an Egyptian MiG-19 was shot down. In the Six-day War, all 58 victories claimed by the Israelis were with guns.

In fact, the Israeli pilots used the classic slashing attacks wherever possible, and avoided turning in most cases. There is one recorded instance where a Mirage squadron commander fought a Jordanian Hunter for no less than eight and a half minutes before he could administer the *coup de grâce:* this must be one of the longest all-jet combats on record, and speaks volumes for the inability of the Mirage to turn with the subsonic Hunter.

On the other hand, the speed loss attendant upon hard turns could be used to advantage to force an overshoot. The Armée de l'Air taught a sudden, vicious pitch-up as an evasion manoeuvre; the tremendous speed loss almost invariably forced an attacker to overshoot, but it left the defender out of airspeed and vulnerable to a further attack. No firm figures are available, but a close approximation of the war record of the Mirage III is about 200 victories (something over half with missiles) to about two dozen losses in the air-to-air arena. A British Fleet Air Arm pilot who had exercised extensively against the French commented: 'They just screamed out from behind a cloud, or out of the sun, into a firing pass. They wouldn't even deign to so much as drop a wing for us.' Tactically, this was perfectly correct.

The next major variant after the IIIC was the IIIE, which was slightly longer, a bit heavier, more powerful, with more internal fuel, and with a better avionics system, mainly to improve the air to ground ordnance delivery. Most Mirage IIIs can reach 40,000ft (12,200m) in about 6.5 minutes from brake release, although

Above: Three Mirage IIIEs of the Armée de l'Air carry (right to left) air-to-ground, air-to-air and anti-shipping weapons, plus drop tanks.

the Mirage IIICZ operated by South Africa is fitted with the considerably more powerful Atar 9K50 and can almost certainly do better. Using the SEPR rocket motor gives a time to 60,000ft (18,300m) of 7 minutes 20 seconds, and also increases the ceiling to 75,500ft (23,000m).

Israeli experience in the clear skies of the Middle East led to the development of the Mirage 5, which first flew in 1967. This version, shorn of much of the adverse weather avionic system, was rather lighter than the IIIE but carried more fuel. In essence it was a ground-attack Mirage which could operate as a clear weather fighter. Political problems ensured that Israel never received its Mirage ▶

▶ 5s, but some of them found their way to Egypt during the October War of 1973, and Mirage III clashed with Mirage 5, a circumstance which led the Israelis to adopt black-edged yellow or orange triangle markings on their Mirages, a move more than justified by the number of own goals scored during this war.

The Mirage 5 was followed in 1979 by the Mirage 50, a 5 hotted up by the installation of the Atar 9K50. Some IIIs were also fitted with this engine and designated 50s. The Mirage 5 had a very similar performance to the III, but the more powerful Mirage 50 had a considerably better initial rate of climb and acceleration, and could reach 40,000ft (12,200m) from a standing start in under 5 minutes. Take-off distance was also considerably improved.

Sales of the Mirage 5 were quite respectable, but the Mirage 50 attracted very little interest. A considerably revised aircraft, the Mirage 3NG, is powered by the Atar 9K50; it has a new avionic fit and fly-by-wire is used in the control system. The delta planform has been modified with a small leading edge extension at the wing root and canard foreplanes which, coupled with FBW and artificial stability, greatly improve manoeuvrability.

Canards on a Mirage delta are not a new idea: the Israeli Aircraft Industries Kfir, derived from the Mirage III, has featured canard foreplanes since 1973, while Dassault's Milan flew in September 1968 with small retractable foreplanes, though neither featured relaxed static stability coupled with FBW. To date, the Mirage IIING has no takers, and it is unclear whether the NG is being offered as a new-build aircraft or in the form of a mid-life update.

The Swiss operate a purpose built Mirage, the IIIS, which has much the same performance as the IIIE but costs twice as much. They are currently planning an update which features nose strakes, canards, a better ejection seat and more modern avionics, while a wing strengthening programme is proposed to extend aircraft life into the next century. The strakes are to improve lateral stability at high angles of incidence, while the canards add lift forward of the centre of gravity. A Swiss Air Force statement credits the canards with increasing the instantaneous turn rate to match that of the F-16 but with less speed loss.

Chile is also planning to modify its fleet of Mirage 50s with canards with the help of IAI, whose package offers an increased sustained turn rate of 19°/sec (from 14.5°/sec), with an

Below: Canard foreplanes can be used to offset some of the disadvantages of the tailless delta layout. This is the sole example of the Mirage 3NG: to date there have been no sales.

improvement in angle of incidence allowable from 27° to 42°, while Atlas Industries are undertaking a similar midlife update on South Africa's ageing force of Mirage IIIs under the new name of Cheetah.

The use of canards generally approximately doubles the lift:drag ratio and greatly improves low speed manoeuvrability. Vmin is claimed to reduce from around 150kt (278km/hr) to 107kt (198km/hr), which would bring about a reduction in corner velocity from, in rough terms, 400kt (740km/hr) to 300kt (556km/hr).

The final improvement proposed for the Mirage III family is an in-flight refuelling capability. Dassault-Breguet have developed an upgrade which involves the installation of a new nose bay ahead of the cockpit, involving a 3.6in (9cm) increase in length. The probe, which can be unbolted but is not retractable, is offset to the right ahead of the cockpit. A buddy pack can be used for operational refuelling.

Above: A Belgian Mirage 5 with large fuel/multi-sensor pods. Originally designed as a clear air attack fighter, the Mirage 5 has a slimmer nose and more austere avionics than the III.

Armament
Early versions: one R.530 on the centreline
Most versions: two underwing R.550s or Sidewinders
All versions (except when carrying rocket motor and fuel): two 30mm DEFA cannon with 125 rounds per gun

Users
Argentina, Australia, Belgium, Brazil, Chile, Colombia, Egypt, France, Gabon, Israel, Libya, Pakistan, Peru, South Africa, Spain, Switzerland, United Arab Emirates Venezuela, Zaire. Lebanon has some Mirages but cannot be said to operate them, while Colombia's Mirage 5s are being fitted with canards and the J79-GE engine.

McDonnell Douglas F-4 Phantom

Type: Two-seat all-weather carrier-based interceptor, adapted as land-based interceptor, air superiority fighter, and fighter-bomber; reconnaissance and Wild Weasel variants exist.

It is not surprising that the versatility of the Phantom has become a byword. At the design stage it was really no more than a series of solutions looking for the right problem. At the time, fast jet bombers were coming into service armed with long-range anti-ship missiles, and the US Navy, increasingly conscious of its global role, needed to protect its giant carriers at long ranges.

The McDonnell F-4 Phantom was accordingly developed as a Fleet Air Defence Fighter able to launch, cruise out to 250nm (463km) from the carrier, stay on patrol at that distance, detect and intercept intruders with missiles from beyond visual range if necessary, and return to the carrier three hours later. It was to be fast enough to catch any existing or projected threat aircraft, but the ability to outfight it when caught did not arise; the colossal speeds of the new breed of fighters, and the

Dimensions	F-4E	F-4F	F-4J/F-4S
Length (ft/m)	63.00/19.20	63.00/19.20	58.16/17.73
Span (ft/m)	38.33/11.68	38.33/11.68	38.33/11.68
Height (ft/m)	16.25/ 4.95	16.25/ 4.95	16.25/ 4.95
Wing area (sq ft/m²)	530/49.25	530/49.25	530/49.25
Aspect ratio	2.77	2.77	2.77
Weights			
Empty (lb/kg)	29,535/13,400	28,400/12,880	28,000/12,700
Takeoff (lb/kg)	45,750/20,750	42,000/19,050	44,220/20,060
Combat (lb/kg)	39,250/17,800	35,800/16,240	37,710/17,100
Power	2xJ79-17	2xJ79-17	2xJ79-10
Max (lb st/kN)	17,900/79.6	17,900/79.6	17,900/79.6
Mil (lb st/kN)	11,870/52.8	11,870/52.8	11,870/52.8
Fuel			
Internal (lb/kg)	13,020/5,900	12,400/5,625	13,020/5,900
External (lb/kg)	Variable	Variable	Variable
Fraction	0.28	0.30	0.31
Loadings			
Max thrust	0.78 − 0.91	0.85 − 1.00	0.81 − 0.95
Mil thrust	0.52 − 0.60	0.57 − 0.66	0.54 − 0.63
Wing takeoff (lb/sq ft/kg/m²)	86/421	79/387	83/407
Wing combat (lb/sq ft/kg/m²)	74/361	68/330	71/347
Performance			
Vmax hi	M = 2.2	M = 2.2	M = 2.2
Vmax lo	M = 1.19	M = 1.19	M = 1.19
Ceiling (ft/m)	55,000/16,750	55,000/16,750	55,000/16,750
Initial climb (ft/min/m/sec)	28,000/142	28,000/142	28,000/142
Takeoff roll (ft/m)	3,300/1,000	N/A	N/A
Landing roll (ft/m)	N/A	N/A	N/A
First flight	Aug 1965	May 1973	Jun 1965

new wonder missiles had, so it was said, rendered the dogfight obsolete.

A two-man crew was needed to handle the advanced radar and weapon system, and two engines were needed to provide the required level of performance for what was a large fighter. Armament was four of the new semi-active radar homing Sparrow missiles, although provision was made for two more Sparrows or four heat-homing Sidewinders to be carried beneath the wings.

The first major variant to see service was the F-4B, but with the Phantom taking nearly every performance record in sight, Air Force interest was aroused, and intensive evaluation led to the F-4C and F-4D being ordered for the USAF. Both Air Force and Navy were heavily involved in the Vietnam War, and it was here that the

Phantom first showed its versatility, gradually taking over almost every role from almost every other type on the inventory.

So far as can be ascertained, the Phantom has notched up more aerial victories than any other type still in front-line service. These have been scored in two main theatres: South East Asia in United States service, and in the Middle East in Israeli service. Iranian Phantoms are still in action in the protracted war with Iraq, but no firm and reliable information is available, except that two have been shot down by Saudi Arabian F-15s.

At the last count, the score was 283, the majority of victims being Soviet designs, and so far as is known these comprise 53½ MiG-17s, 10 MiG-19s, 109½ MiG-21s, and two An-2 transports. The rest, with ▶

F-4M (FGR.2)
58.75/17.91
38.33/11.68
16.25/ 4.95
530/49.25
2.77

| 31,000/14,060 |
| 47,220/21,420 |
| 40,710/18,470 |

| 2xSpey 203 |
| 20,515/91.2 |
| 12,250/54.5 |

| 13,020/5,900 |
| Variable |
| 0.28 |

| 0.87 — 1.01 |
| 0.52 — 0.60 |
| 89/435 |
| 77/375 |

| M = 2.1 |
| M = 1.2 |
| 55,000/16,750 |
| 32,000/163 |
| N/A |
| N/A |

| Feb 1967 |

Below: F-4Bs of US Navy fighter squadron VF-121 photographed in December 1967. Developed as a carrier-based interceptor, the Phantom has fulfilled many other roles since.

▶ one exception, are identified as MiGs, and are victims of Israeli Phantoms. The one exception was a British Jaguar shot down over Germany in error. The weapons used in these victories vary. The SARH Sparrow has accounted for 64, the IR-homing Sidewinder and Shafrir 97½, the IR Falcon 5, the 20mm cannon 22½, and manoeuvring — either luring the opponent past the point of lost control, or running him out of fuel — is responsible for five, while the others, mainly Israeli again, remain unknown.

The Phantom's first kills were scored by US Navy F-4Bs over Vietnam in 1965. It was, however, quickly found that BVR identification was a major problem, and this restricted the use of the Sparrow to a great degree. After a couple of own goals visual identification became the order of the day except in very exceptional circumstances, so the Phantom became involved in the close combat arena, for which it had never been designed.

Even when BVR combat was permitted under the Rules of Engagement, it was found that the small MiGs were difficult to hold on radar for missile guidance. Nor did the missiles do anywhere near as well as they had during trials; they had been designed to cope with non-manoeuvring targets, and the North Vietnamese Air Force were not so cooperative. Furthermore, the missiles had been designed to be launched from fighters pulling no more than 2½ g, a limitation that did not fit well with close combat.

The avionic and weapon system of the Phantom was better than anything possessed by the NVAF MiGs, but was at least offset by the comprehensive GCI environment set up in North Vietnam. In close combat the large and smoky engined Phantom could be seen and identified at much greater distances than its small opponents, which could often approach undetected. The Phantom lacked agility in the lower speed range and, worst of all, it lacked a

gun, so that it was often caught up in a dogfight without a close-range weapon. Finally, it could be difficult to fly, suffering from dihedral effect, and had rather unforgiving departure tendencies, which accounted for many losses. An inexperienced pilot tended to spend too much time flying it and not enough time fighting.

The early Phantoms to see combat were the US Navy and Marine Corps F-4B and the Air Force F-4C. Operational needs and technical progress quickly led on to new variants, the first of which was the F-4J for the Navy, with more powerful engines, increased fuel capacity, and improved avionics. This was quickly followed by the USAF F-4D with better ground-attack capability. A few F-4Js and Ds remain in service even today, but not many.

The greatest advance came with the USAF F-4E which was developed in parallel with the F-4J to give greater air-to-air capability, with an improved radar and weapons control system, an internal 20mm cannon,

and slats on the wing leading edges to improve manoeuvring. This variant was also used in combat by Israel from the War of Attrition onwards.

Next came two specially developed Phantoms for the Royal Navy and Royal Air Force respectively, the F-4K and F-4M, known in British service as the FG.1 and FGR.2. Neither had an internal gun, but both could, like the earlier Phantoms, carry gun pods, though these were inaccurate during manoeuvres as a result of mounting distortion. Nor did they feature the wing slats of the F-4E. The greatest difference was that they were powered by Rolls-Royce Spey turbofans, which in theory were to give improved climb, acceleration, fuel consumption and low-speed handling. Problems were encountered in matching the engines to ▶

Below: Eight missiles against two, longer range and superior radar were the Phantom's main advantages over the Lightning.

Above: This F-4E of the 388th TFW, flying from Korat against North Vietnam in 1972, has the original short gun muzzle.

Below: Israeli F-4Es have seen a great deal of action in the Middle East, although their main role is ground attack.

the inlets, and many of the anticipated advantages never materialised.

Apart from specialised reconnaissance variants, and the F-4G Wild Weasel, the final Phantom to be developed was the F-4F for the Luftwaffe, a dedicated air superiority fighter with the slatted wing of the E, the internal cannon, various weight-saving measures to lower wing loading and increase thrust loading, and an armament of just four Sidewinders. At other times, F-4Bs were upgraded into F-4Ns, and F-4Js into F-4Ss.

More than 5,000 Phantoms were built, and it is estimated that about 2,000 of them will still be in service at the turn of the century. There is, therefore, a flourishing market in mid-life updates for Phantoms. The US Navy phased out its last F-4s in 1986, although the Marine Corps retains a considerable number. The USAF has a few F-4Ds and quite a lot of F-4Es that it intends to upgrade.

The Royal Navy relinquished its F-4Ks to the Royal Air Force, where they still equip two squadrons, but are fast running out of hours, but the F-4Ms will be around until well into the 1990s. Many other nations have low-mileage Phantoms in service, and given the enormous cost of new hardware will be looking to upgrade them.

The most comprehensive upgrade proposals put forward to date, those of the Boeing Military Airplane Company in conjunction with Pratt & Whitney Aircraft, involve re-engining, a new avionics pack and a conformal fuel tank. The suggested engines, also favoured by Israel, are PW1120s giving 15 per cent more thrust along with reduced weight and fuel consumption. The avionics updates mainly concern the fitting of modern multi-mode radars such as the Hughes APG-65, which will confer a tremendous increase in capability. Structural strengthening is also a possibility, either to increase the gross weight limit, or lengthen the structural life.

To date three update programmes are under way; Japan is going for extended airframe life coupled with advanced fire control and avionics; West Germany is having an Improved Combat Efficiency (ICE) avionics pack, including APG-65, and compatibility with AIM-120 Amraam when it becomes available; and Israel seems to favour a full house, with the PW.1120 engine, modern avionics, life extension, and possibly canard foreplanes to improve manoeuvrability and handling.

Armament
Almost all: four AIM-7 Sparrows and four AIM-9 Sidewinders
West German F-4F: four Sidewinders (four Amraam in future)
British: four Sky Flash as an alternative to Sparrow
Israel: four Shafrir as alternative to Sidewinder
All F-4E and F-4F have internal M61 20mm cannon; all others can carry a 20mm gun pod on centreline

Users
USA, UK, West Germany, Israel, Iran, South Korea, Spain, Japan, Greece, Turkey

Northrop F-5E Tiger II and F-20 Tigershark

Type: Single-seat air superiority fighter/quick-reaction interceptor with some adverse weather capability, some ground attack capability. Dedicated reconnaissance and two-seater trainer exist.

Unusually, the F-5 series owes its origins to a trainer, the T-38 Talon, which was developed to meet a US Air Force requirement for a supersonic advanced trainer. Northrop decided to develop a fighter version as a private venture; the result was the F-5A, a small, austere fighter which had the advantage of being cheap, yet supersonic. In April 1962 it was adopted by the US Department of Defense as the Military Assistance Program all-purpose fighter, which meant that it would be supplied to friendly countries on very advantageous terms. Simple and easy to maintain, the Freedom Fighter, as the F-5A was named, could do a convincing job against a basic air threat, but lacked the payload/range capability to pose a threat outside the national borders, having what was once pithily described as 'the ability to carry a box of matches the length of a baseball park'.

Dimensions	F-5E Tiger II	F-20 Tigershark
Length (ft/m)	47.39/14.44	53.92/16.44
Span (ft/m)	26.19/ 7.98	26.19/ 7.98
Height (ft/m)	13.34/ 4.07	13.75/ 4.19
Wing area (sq ft/m²)	186/17.28	186/17.28
Aspect ratio	3.69	3.69

Weights		
Empty (lb/kg)	9,683/4,392	13,376/6,070
Takeoff (lb/kg)	14,847/ 6,735	18,540/8,510
Combat (lb/kg)	12,665/ 5,745	16,015/7,265

Power	2xJ85-21A	1xF404-100
Max (lb st/kN)	5,000/22.2	18,000/80.0
Mil (lb st/kN)	3,600/16.0	12,000/53.3

Fuel		
Internal (lb/kg)	4,364/1,980	4,364/1,980
External (lb/kg)	4,466/2,025	5,363/2,430
Fraction	0.29	0.24

Loadings		
Max thrust	0.67 — 0.79	0.97 — 1.12
Mil thrust	0.48 — 0.57	0.65 — 0.75
Wing takeoff (lb/sq ft/kg/m²)	80/390	100/487
Wing combat (lb/sq ft/kg/m²)	68/332	86/420

Performance		
Vmax hi	M = 1.64	M = 2.0
Vmax lo	M = 0.95	M = 1.2
Ceiling (ft/m)	51,200/15,600	55,000/16,750
Initial climb (ft/min/m/sec)	34,300/174	53,800/273
Takeoff roll (ft/m)	2,000/610	1,475/450
Landing roll (ft/m)	N/A	N/A

First flight	Aug 1972	Aug 1982

The next step was the more powerful and better performing F-5E Tiger II which proved itself a dangerous opponent in simulated close combat, being selected to equip the USAF Aggressor squadrons. In subsonic performance and manoeuvrability it was closely comparable to the MiG-21, and it had a similar armament of two 20mm cannon in the nose and two Sidewinders carried on wingtip rails. Like its Soviet counterpart, the F-5E is small, has smokeless engines and is difficult to see. Turn rates at Mach 0.9 and 15,000ft (4,600m) are 14°/sec instantaneous and 7.8°/sec sustained; at Mach 0.5 at the same altitude instantaneous turn is 11.4°/sec. The cg is too far forward, which makes it a trifle sluggish in pitch, but its sparkling rate of roll more than compensates.

When it was realised that Mach 2 was irrelevant in almost every combat situation, and that manoeuvrability was still important, many nations bought the F-5E, which was, unlike many other fighters, affordable both in acquisition and in running costs. The obvious next step was an even better version, which emerged in the summer of 1982 as the F-5G.

The F-5G, or F-20 Tigershark as it later became, differed from the F-5E primarily in being powered by a single General Electric F404 turbofan. The avionics fit was updated to suit the requirements of the 1990s, with a modern multi-mode radar, and performance benefited from the increase in thrust loading. The new fighter was stressed for 9g turns: 19.4°/sec instantaneous turn has been demonstrated, and 13.1°/sec ▶

Above: Big missiles on small fighters never look right, as this F-20 Tigershark armed with Sparrows demonstrates.

Below: The F-20, an upgraded version of the F-5E Tiger II, is an outstanding fighter in the close combat arena.

▶ sustained at Mach 0.8 and 15,000ft (4,600m). A time of just 2.3 minutes to 40,000ft (12,200m) from a standing start has also been recorded, and an acceleration of 14kt (26km/hr) at 10,000ft (3,050m) is attainable.

As a quick-reaction air superiority fighter at a reasonable price, the Tigershark looks unbeatable, yet so far there are no takers. In August 1986 the fourth prototype was under construction, and this will have even better performance than the first three, coupled with increased manoeuvrability. Tests have been carried out using Sparrow, but carrying such a big missile on such a small fighter must be like putting an anchor on it.

Below: The Tigershark is also advertised as an interceptor. It can be airborne, with all systems fully up, in just 60 seconds from engine start, a feat unmatched by other types.

The F-20 is currently in competition with the F-16 for a 300-aircraft USAF continental air defence purchase, the result of which was to be announced in November 1986; it has also been checked out with many air-to-ground weapons as a contender to replace the A-10. Many countries have shown interest, among them South Korea, Switzerland, Taiwan and Jordan, and Oman actually tried to place an order for six in 1985, but that was too small for production to be initiated.

The F-20 is an object lesson in how unpredictable the fighter market can be. The F-5A was a success, as was the F-5E, while the F-18, originally intended as an F-5E replacement, ended up as a very expensive multi-role fighter, beyond the purse of most F-5E operators. Against this background the Tigershark looked certain to scoop the light fighter market, but as yet it hasn't.

Weapons

F-5E and F-20: two 20mm M39 revolver cannon with 225 rounds per gun and two Sidewinders on the wingtips

F-20 only: two AIM-7F Sparrow demonstrated, Amraam and Asraam in the future

Users

(F-5E/F only) Bahrain, Brazil, Chile, Ethiopia, Indonesia, Iran, Jordan, Kenya, South Korea, Malaysia, Mexico, Morocco, Saudi Arabia, Singapore, Sudan, Switzerland, Taiwan, Thailand, Tunisia, USA, South Yemen. A total of 19 nations have operated or are still operating the F-5A Freedom Fighter.

Right: The F-5E Tiger II is the most widely used Western fighter. Head-on it shows only a small profile, which is all but invisible at 2nm (4km).

Mikoyan MiG-25 Foxbat

Type: Single-seat point defence interceptor with adverse weather capability; intercept automated through data link via GCI. Reconnaissance and two-seat trainer versions exist.

The MiG-25 Foxbat is unique, being the only fighter designed and built to counter a single-threat aircraft. It is also unique in that, reconnaissance apart, it has no other role than the interception of fast and high intruders. It was originally conceived as a counter to the US Air Force B-70 Valkyrie, a nuclear-armed bomber designed to fly the entire mission at Mach 3 and 80,000ft (24,400m).

It hardly needs saying that the in-terception of an aircraft flying at around a statute mile every two seconds (nearly 1km/sec) some 15 statute miles (24.4km) up in the stratosphere, is a very tall order and demands precision of a high degree. It did not seem likely that the surface-to-air missiles then under development would be able to achieve satisfactory results, and the Soviet solution was to design an aircraft optimised for the highest speeds and altitudes, with a rapid rate of climb, the ability to carry large missiles, and a ground control system of sufficient precision to position it where its own radar could be used for the final phase of the intercept. At the same time, the geography of the Soviet

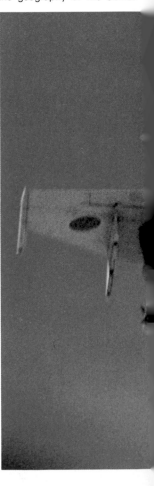

Dimensions	MiG-25 Foxbat-A
Length (ft/m)	71.33/21.74
Span (ft/m)	46.00/14.02
Height (ft/m)	18.50/ 5.64
Wing area (sq ft/m²)	662/61.52
Aspect ratio	3.20

Weights	
Empty (lb/kg)	44,000/19,960
Takeoff (lb/kg)	82,500/37,420
Combat (lb/kg)	66,710/30,260

Power	2xR.31
Max (lb st/kN)	27,120/130.5
Mil (lb st/kN)	20,500/ 91.0

Fuel	
Internal (lb/kg)	31,575/14,320
External (lb/kg)	nil
Fraction	0.38

Loadings	
Max thrust	0.66 — 0.81
Mil thrust	0.50 — 0.61
Wing takeoff (lb/sq ft/kg/m²)	125/608
Wing combat (lb/sq ft/kg/m²)	101/492

Performance	
Vmax hi	M = 2.82
Vmax lo	M = 0.85
Ceiling (ft/m)	78,000/23,800
Initial climb (ft/min/m/sec)	41,000/208
Takeoff roll (ft/m)	4,528/1,380
Landing roll (ft/m)	7,152/2,180

First flight	Aug 1964

Union meant it needed to be affordable in reasonable numbers.

Foxbat finally emerged as a manned re-usable first stage of a missile system for all practical purposes. It had to be automated to a high degree: extreme relative speeds would be involved in interception, and visual acquisition could play little part, since at its operational altitudes the sky is deep blue, and the sun blazes like a giant star, so that unless the target is outlined against the earth below, there is little chance of it being seen by an interceptor.

The key to interception lay in ground control coupled with a digital data link tied to an automatic flight control system, designed to control the aircraft from just after takeoff to just before touch down. As with the American F-106, the pilot was basic-ally a systems manager, responsible for takeoff and landing, throttle control, selection of reheat, missile launch and the post-attack break. Of course, if circumstances demanded he could fly the mission manually with GCI advisory information, but that would only be done if the basic auto system was inoperable.

Foxbat is big. The destruction of very high and fast targets demands big missiles and four were carried in the form of the AA-6 Acrid, two SARH versions on the outer underwing pylons and two IR homers inboard. Target acquisition was the task of the large and extremely powerful Foxfire radar, with a range ▶

Below: MiG-25 Foxbat-A of the Libyan Arab Air Force armed with two AA-6 Acrid missiles.

against a Valkyrie-sized target of about 50nm (93km) and enough power to burn through any jamming carried by Valkyrie. The standard attack was either collision course from the front quarter, or from the beam: Foxbat would have stood little chance in a tail chase; it could be wound up to Mach 3.2, but only at the cost of wrecking its engines through overspeeding.

It was clear at the design stage that the MiG-25 would be forced to use prolonged spells of afterburning to achieve the required levels of peformance, and that this would demand huge amounts of fuel. The fuel fraction of Foxbat is high at 0.38, and the actual amount carried is enormous, more than 14 tonnes. Even so, the full-burner interception radius is a mere 160nm (296km), since to counter the effects of kinetic heating, the aircraft is constructed with large amounts of heavy nickel steel.

Manoeuvre combat never figured in the design of the MiG-25. With full internal fuel it is limited to just over

Above: The Foxbat-E seen here carrying four AA-6 missiles is an upgraded Foxbat-A, with High Lark radar and a revised nose. The pilot's view from the cockpit is very restricted.

2g, while with more than half the fuel burned, its limitation is still only 5g — much less than any other fighter in service. And, reconnaissance variants apart, its service record is not impressive. To date, three Syrian Foxbat-As have fallen to the Israeli F-15/Sparrow combination.

Most Foxbat-As have been upgraded to Foxbat-E standard, with improved avionics including the High Lark radar, as used on the MiG-23, which has a much better performance at low level, coupled with the AA-7 Apex missile, carried in SARH and IR homing pairs. Foxbat-E is rumoured to have been structurally strengthened to allow manoeuvre combat, but it seems likely that any radical change would involve a severe weight penalty.

Armament
Four AA-6 Acrid or AA-7 Apex missiles (2 SARH and 2 IR homing)

Users
Algeria, Libya, Syria, USSR

Below: MiG-25R reconnaissance Foxbats. The B in the foreground carries cameras while the D behind it is equipped with electronic sensors and side-looking airborne radar.

Shenyang F-7 and F-7M Airguard

Type: Single-seat lightweight point defence and close combat fighter with limited all-weather and attack capability.

The Shenyang F-7 (J-7 in Chinese service) was basically an unlicensed version of the MiG-21F with similar characteristics, but only three or four score were built before production ceased from 1966 until the late 1970s, when production resumed of an improved version with a more reliable engine, greater endurance — the fuel fraction is well up on that of the original MiG-21C — and two 30mm cannon instead of the previous single NR-30.

The F-7M Airguard is essentially a revised F-7, rather sleeker in apperance than most MiG-21s, with Western avionics, including a ranging radar and HUDWAC. Details were first released in October 1984. Takeoff gross weight is, at 19,620lb (8,900kg), rather more than that of the MiG-21C but less than that of later models. It has five hardpoints, but only the two inboard wing positions appear to be used for AAMs; the centreline position seems to carry nothing but a fuel tank and the two

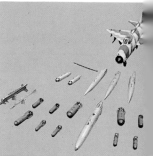

Dimensions	F-7M Airguard
Length (ft/m)	45.75/13.95
Span (ft/m)	23.46/ 7.15
Height (ft/m)	13.45/ 4.10
Wing area (sq ft/m²)	247/23.00
Aspect ratio	2.23

Weights	
Empty (lb/kg)	11,630/5,275
Takeoff (lb/kg)	17,400/7,900
Combat (lb/kg)	14,900/6,760

Power	1xWP-7BM
Max (lb st/kN)	13,448/60.0
Mil (lb st/kN)	7,540/33.5

Fuel	
Internal (lb/kg)	5,000/2,270
External (lb/kg)	2,575/1,168
Fraction	0.29

Loadings	
Max thrust	0.78 — 0.91
Mil thrust	0.45 — 0.53
Wing takeoff (lb/sq ft/kg/m²)	70/344
Wing combat (lb/sq ft/kg/m²)	60/295

Performance	
Vmax hi	M = 2.05
Vmax lo	M = 1.01
Ceiling (ft/m)	59,700/18,200
Initial climb (ft/min/m/sec)	35,400/180
Takeoff roll (ft/m)	3,100/950
Landing roll (ft/m)	2,950/900

First flight	Dec 1964

outboard wing positions are also plumbed for fuel.

Basically a clear air fighter, Airguard is capable of 40 minutes on CAP with a five-minute combat fuel allowance at a distance of 54nm (100km) from base at 36,000ft (11,000m) while carrying two AAMs and three 500lit drop tanks. With the same configuration it can carry out a high level supersonic intercept at a maximum distance of 350nm (650km).

Widely touted on the export market, Airguard offers a cheap form of defence against a limited threat and looks likely to sell well. Pakistan has evaluated it as a replacement for its remaining F-6s and is believed to have ordered between 60 and 150 aircraft. This order was won in competiton with the Mirage 3NG and F-20

Tigershark, and while it is not suggested that the Airguard is superior to either of those, it may well be a better buy in terms of equal cost quantities where a simple mission and numerical strength are the main factors. Much modern thinking seems to indicate that a cheap airframe fitted with Western avionics is an adequate compromise, and the Airguard conforms to this trend, although it will be too basic for many tastes.

Armament
Two 30mm cannon with 60 rounds per gun; 2 PL-2, PL-2A, or PL-7 missiles. Airguard is also compatible with Sidewinder and Magic.

Users
Albania, China, Egypt, Iraq, Pakistan, Tanzania, ZImbabwe

Left: the F-7M Airguard can carry a variety of weapons, both air-to-air and air-to-ground, as depicted here.

Below: The F-7M is a direct descendant of the MiG-21. With Western avionics it is a cost-effective light fighter.

Sukhoi Su-15/-21 Flagon

Type: Single-seat adverse weather interceptor with secondary air superiority role. Two-seat trainer versions exist.

The Su-15 Flagon was the Soviet Union's first really effective all-weather interceptor, taking over the role from the Su-9/11 family. Unusual for a Soviet fighter in having a high wing loading, with a consequent lack of manoeuvrability, and being reliant on long concrete runways, the Su-15 was optimised for a high rate of climb and high speed and featured an automated interception system, making it a Soviet counterpart of the American F-106, although it did not carry missiles internally. Flagon-D,

the first model to see service in large numbers, featured an extended wing span with a kinked leading edge.

In 1973 an uprated version, Flagon-E, entered service with a better radar and avionics, the detection range against fighter-sized targets being of the order of 43nm (80km), and more powerful engines; it differed sufficiently from earlier models to warrant the new designation Su-21, and has since been followed by Flagon-F.

The Flagon is a big fighter, but very aerodynamically clean, its hardly outstanding thrust loading propelling it to 36,000ft (11,000m) from a standing start in 2½ minutes. Its fuel fraction is rather low, despite some fuel being housed in the fin, but its com-

Dimensions	Su-21 Flagon-F
Length (ft/m)	72.00/21.95
Span (ft/m)	34.50/10.52
Height (ft/m)	16.50/ 5.03
Wing area (sq ft/m²)	385/35.78
Aspect ratio	3.09

Weights	
Empty (lb/kg)	27,000/12,250
Takeoff (lb/kg)	41,400/18,780
Combat (lb/kg)	35,900/16,285

Power	2xR-13F2-300
Max (lb st/kN)	15,875/70.5
Mil (lb st/kN)	11,240/50.0

Fuel	
Internal (lb/kg)	11,000/4,990
External (lb/kg)	2,750/1,250
Fraction	0.27

Loadings	
Max thrust	0.77 — 0.88
Mil thrust	0.54 — 0.63
Wing takeoff (lb/sq ft/kg/m²)	108/525
Wing combat (lb/sq ft/kg/m²)	93/455

Performance	
Vmax hi	M = 2.05
Vmax lo	M = 1.10
Ceiling (ft/m)	65,600/20,000
Initial climb (ft/min/m/sec)	45,000/229
Takeoff roll (ft/m)	'long'
Landing roll (ft/m)	N/A

First flight	1964 (Flagon-E)

bat radius on internal fuel only is quite respectable, being stated as 390nm (725km), although this is certainly not a full-burner, flat-out interception. Some sources state that Flagon-F is powered by Tumansky R-25 turbojets, which would bring the thrust:weight ratio at combat weight to well over unity and improve non-afterburning performance substantially, though probably at the expense of range.

Flagon has served only with the USSR, being too limited and probably too expensive for the sort of Third World countries that Soviet fighters are usually exported to, while it needs an integrated ground control system to be really effective. So far as is known it does not carry a gun, although there have been rumours of the twin-barrel GSh-23 being mounted in pods. And it is not likely to be of much use in close combat; an approximate calculation puts corner velocity at around 470kt (870km/hr), which would mean a large turn radius and a slow turn rate. The effect of this can be judged by comparison with the MiG-21 with a V corner of around 300kt (556km/hr), which is judged to be quite agile. At present, only the Su-21 is in service, in the form of Flagon E and F.

Armament
Four AA-3-2 Advanced Anab (two SARH and two IRH); possibly AA-7 Apex and AA-8 Aphid

Below: The clean lines of the Su-21 Flagon-E belie its great size. The missiles carried are AA-3 Advanced Anabs.

Mikoyan MiG-23 Flogger

Type: Single-seat multi-role fighter. All-weather interceptor, specialist ground-attack and two-seat trainer versions exist.

The MiG-23 Flogger can fairly be described as the first Soviet attempt to produce an all-round tactical fighter with a useful payload/range capability. It is also the first Soviet variable-geometry aircraft to enter service. The basic requirements were to be able to match the American Phantom in the air-to-air arena, both in performance and in weapon system, which meant a comparative-ly long-range detection and tracking radar coupled with missiles that could be launched from beyond visual distance. It also called for an operational radius far exceeding that of the MiG-21, combined with high speed and good rate of climb.

While manoeuvrability is nice to have in close combat, surprise is, was and always has been the dominant factor in air combat: performance was needed to achieve surprise, and the manoeuvrability requirement was distinctly secondary. Short field capability was to remain the same as that of the MiG-21. Finally, as with most Soviet warplanes, it was to be cheap and easy to maintain.

The advantages of variable wing sweep are threefold. Firstly, slower take off and landing speeds can be used, with a consequent shortening

	MiG-23MF Flogger-G
Dimensions	
Length (ft/m)	55.50/16.92
Span (ft/m)	46.75/14.25 max
Height (ft/m)	14.33/ 4.37
Wing area (sq ft/m²)	325/30.20
Aspect ratio	6.72 − ˙2.27
Weights	
Empty (lb/kg)	25,000/11,340
Takeoff (lb/kg)	38,000/17,240
Combat (lb/kg)	33,043/14,990
Power	1xR-29B.
Max (lb st/kN)	25,350/112.7
Mil (lb st/kN)	17,635/ 78.4
Fuel	
Internal (lb/kg)	9,914/4,500
External (lb/kg)	4,100/1,860
Fraction	0.26
Loadings	
Max thrust	0.67 − 0.77
Mil thrust	0.46 − 0.53
Wing takeoff (lb/sq ft/kg/m²)	117/571
Wing combat (lb/sq ft/kg/m²)	102/496
Performance	
Vmax hi	M = 2.35
Vmax lo	M = 1.10
Ceiling (ft/m)	55,000/16,750
Initial climb (ft/min/m/sec)	N/A
Takeoff roll (ft/m)	2,953/900
Landing roll (ft/m)	N/A
First flight	circa 1965

of necessary runway length. Secondly, endurance, or loiter time, can be increased. And thirdly, the wings can be reconfigured for the optimum manoeuvre, high speed, or low altitude requirement. Flogger has three wing sweep settings, 16°, 45° and 72°, set manually, and unlike Western VG fighters, the minimum setting is not combat rated; short field performance was the driving factor, not manoeuvre capability.

Two MiG-23 variants have been configured for air combat, the MiG-23M Flogger-B, which entered service in the early 1970s, and the MiG-23MF Flogger-G which is the current edition. Performance data and general remarks passed on Flogger performance are conflicting: some sources ascribe an initial rate of climb exceeding 50,000ft/min (254m/sec),

while others refer to its 'jackrabbit acceleration', neither of which seem justified by the modest thrust:weight ratio, even though the airframe is particularly clean. Nor is there any agreement on whether the engine is a turbojet or a turbofan; the figures issued to date seem to indicate a turbojet. Instantaneous turn rate is nothing special, 11.5°/sec at Mach 0.9 and 15,000ft (4,600m) and 8.6°/sec at Mach 0.5 at the same height. It seems extremely likely that it could be outfought by a well handled Phantom nearly every time.

Flogger-G has a radar codenamed High Lark, which is rumoured to owe ▶

Below: The MiG-23MF Flogger-G will remain the most important Soviet fighter in numerical terms for many years to come.

some of its technology to the AWG-10 carried by the F-4J Phantom. Whatever the truth, High Lark is a pulse-Doppler radar with a search range of 46nm (85km), a tracking range of 30nm (55km) and a limited look down capability. A considerable number of Flogger-Bs carried the Jay Bird radar as fitted to the MiG-21; these were generally Flogger-E export models.

Flogger-G was first seen in the West in 1978 when a small detachment visited Finland and France. The French visit caused a great deal of excitement, with the Soviet fighters having to overfly West Germany on the way to Reims, and every NATO fighter pilot for hundreds of miles was trying to get an air test to see them for himself. It is usually stated that the more sensitive avionics, including High Lark, were stripped out for the trip, but at least two NATO pilots have commented on hearing the distinctive warble of High Lark on their detection kit.

Flogger has seen little combat service to date, and its record is not impressive. Libyan Floggers have been

intercepted over the Mediterranean by US Navy Tomcats on many occasions, and Tomcat crew reaction is that the Libyan Flogger drivers are not as good as those Libyans who fly, say, Mirages. Flogger's only experience of a shooting war has been in Syrian service against Israel over the Beka'a Valley in 1982, when a total of 36 fell to the guns and missiles of Israeli F-15s and F-16s. While failing to score a single victory. It is possible that Iraqi Floggers may have clashed with Iranian fighters, but no firm information is to hand. But however dated Flogger looks, it is important in the fighter world in terms of sheer numbers; in 1986 it was the most numerous fighter in the Soviet inventory.

Armament

MiG-23M and Flogger-E: 23mm GSh-23 with 200 rounds per gun, plus four AA-2 Atolls
MiG-23MF: 23mm GSh-23 with 200 rounds per gun plus two AA-7 Apex and two or four AA-8 Aphid

Users

Algeria, Angola, Bulgaria, Cuba, Czechoslovakia, Ethopia, Finland, East Germany, Hungary, India, Iraq, North Korea, Libya, Poland, Syria, Vietnam, USSR

Left: Unlike Western variable geometry fighters, Flogger is notorious for its lack of turn and manoeuvre capability.

Below: Libya operates the export model Flogger-E, and example of which is seen here with four AA-2 Atolls. Flogger has very poor forward and rear visibility.

Dassault-Breguet Mirage F.1

Type: Single-seat interceptor and air superiority fighter with limited adverse weather capability and considerable ground-attack capability. Reconnaissance and two-seat trainer versions exist.

The Mirage F.1 has little in common with the delta-wing Mirage III series. As we have seen, the delta wing requires very high take off and landing speeds at high angles of incidence, and a correspondingly long runway; it also suffers a very high rate of speed loss due to increased drag in hard manoeuvring, and it is difficult to use high-lift devices to alleviate those problems. A conventional wing will have smaller lifting area with correspondingly high wing loading, but the use of high-lift devices can more than offset that.

In the mid-1960s the Armee de l'Air had a requirement for a big two-seat all-weather multi-role fighter powered by the Pratt & Whitney TF306 turbofan, which emerged as the Mirage F.2. At the same time, Dassault developed a scaled-down single-seat version powered by the Atar, which was preferred to the larger machine, and entered service

Dimensions	Mirage F.1C
Length (ft/m)	50.00/15.24
Span (ft/m)	27.58/ 8.41
Height (ft/m)	14.75/ 4.50
Wing area (sq ft/m²)	269/25.00
Aspect ratio	2.83

Weights	
Empty (lb/kg)	16,315/ 7,400
Takeoff (lb/kg)	25,350/11,500
Combat (lb/kg)	21,658/ 9,824

Power	1xAtar 9K50
Max (lb st/kN)	15,870/70.5
Mil (lb st/kN)	11,060/49.0

Fuel	
Internal (lb/kg)	7,384/3,350
External (lb/kg)	7,540/3,420
Fraction	0.29

Loadings	
Max thrust	0.63 — 0.73
Mil thrust	0.44 — 0.51
Wing takeoff (lb/sq ft/kg/m²)	94/460
Wing combat (lb/sq ft/kg/m²)	80/393

Performance	
Vmax hi	M = 2.2
Vmax lo	M = 1.2
Ceiling (ft/m)	65,000/22,000
Initial climb (ft/min/m/sec)	41,930/213
Takeoff roll (ft/m)	1,968/600
Landing roll (ft/m)	2,198/670

First flight	Dec 1966

Above: The orthodox lines of the
Mirage F.1 tend to obscure the
fact that it outperforms the
Mirage III on all counts.

Below: The single R530 missile
carried on the centreline was
later displaced by two Super 530s
as depicted above.

▶ as the Mirage F.1C. Compared to the Mirage III, its appearance was pedestrian, while its thrust loadings were roughly the same, its wing loadings more than 50 per cent higher and its performance envelope much the same size.

The real difference was made by the wing, which was lavishly equipped with full-span leading edge flaps, and double-slotted trailing edge flaps, greatly increasing lifting capacity, and the difference in performance was dramatic. The approach speed was reduced from 182kt (338km/h) to 139kt (257km/h), with the result that runway requirements were much less, while speed for speed the F.1 could pull about 1g more than the III, with less speed loss in hard turns; the leading edge flaps operate as manoeuvre flaps at high speeds. The empty weight of the F.1 was a little above that of the III, but the fuel fraction was higher and the payload heavier, while both acceleration and initial climb were improved.

The cockpit is fairly basic and the view out can be described as better than the Russians but worse than the Americans. Handling is precise and pleasant, while the ride at high speed and low level is smooth. Using a stepped, maximum thrust climb, the Mirage F.1C can achieve a Mach 2.2 interception 170nm (315km) from base at 40,000ft (12,200m) in less than 12 minutes, with internal fuel only, and armed with guns. This is a trifle unrealistic, and it could reasonably be expected to achieve a Mach 1.8 interception at the same height carrying a full bag of missiles at around 140nm (260km) in the same time, which is still pretty good going.

The radar is Cyrano IVM, a multi-mode radar with a far superior performance to that of the Mirage III's Cyrano II. It has automatic tracking, and limited track-while-scan capability, as well as terrain avoidance and ground mapping. Using the Matra Super 530 missile, it can carry out snap-up attacks on targets well above its own altitude.

The Mirage F.1C-200 is a subtype

fitted with a fixed in-flight refuelling probe to give the French fighter force a long-range deployment capability. Other variants are the F.1A, which is optimised for ground attack, the F.1B two-seat trainer, slightly longer but with less fuel, and the F.1E multi-role strike fighter produced for export.

The Mirage F.1 has seen little air combat to date. It has been widely used by Iraq in the continuing conflict with Iran, but few details have emerged from that war that can be trusted. In South African service the F.1 has downed at least one MiG-21 of the Angolan Air Force.

Armament
Two 30mm DEFA 553 cannon with 135 rounds per gun, two Matra Super 530F SARH missiles, two Matra R550 Magic IR missiles

Users
Ecuador, France, Greece, Iraq, Jordan, Kuwait, Libya, Morocco, Qatar, South Africa, Spain

Above: the Mirage F.1 fulfills both interceptor and counter air roles. The Super 530 AAMs are for BVR use while wingtip Magics are for close combat.

Below: In-flight refuelling probes are fitted to some F.1Cs to give them a long-distance deployment capability.

Saab JA 37 Viggen

Type: Single-seat interceptor and air superiority fighter with adverse weather capability. Fighter, attack, reconnaissance, and two-seat trainer versions each have some capability in a secondary role.

It is hardly possible to build a fighter without it being reconfigured sooner or later for a secondary role. In the case of the JA 37 Viggen, it was intended from the outset that optimised versions of the aircraft would be built for differing roles, and in fact the first Viggen both to fly and to see service was the AJ 37 attack variant; it was several more years before the counter-air JA 37 left the ground.

Sweden is a determinedly unalign-ed country, with a relatively large area and lengthy border to defend. The wonder is that such a numerically small population can afford to design and construct advanced fighters, with a minimum of foreign equipment, in large enough numbers to be convincing as a defence force. Yet the Swedes always seem to manage it: since the end of World War II they have remained entirely self-sufficient in home-brew fast jets. They have made few inroads into the export market, so large sales have not assisted in keeping procurement costs down, and in the case of the Viggen no export sales at all have materialised.

Defending Sweden against air at-

Dimensions	JA 37 Viggen
Length (ft/m)	53.92/16.44
Span (ft/m)	34.77/10.60
Height (ft/m)	18.38/ 5.59
Wing area (sq ft/m²)	495/46.00
Aspect ratio	2.44

Weights	
Empty (lb/kg)	23,650/10,730
Takeoff (lb/kg)	36,000/16,330
Combat (lb/kg)	31,125/14,120

Power	1xRM-8B
Max (lb st/kN)	28,110/125.0
Mil (lb st/kN)	16,200/ 72.0

Fuel	
Internal (lb/kg)	9,750/4,423
External (lb/kg)	N/A
Fraction	0.27

Loadings	
Max thrust	0.78 − 0.90
Mil thrust	0.45 − 0.52
Wing takeoff (lb/sq ft/kg/m²)	73/355
Wing combat (lb/sq ft/kg/m²)	63/307

Performance	
Vmax hi	M = 2 +
Vmax lo	M = 1.1
Ceiling (ft/m)	60,000/18,300
Initial climb (ft/min/m/sec)	40,000/203
Takeoff roll (ft/m)	1,312/400
Landing roll (ft/m)	1,640/500

First flight	Dec 1979

tack and unwanted incursions calls for short reaction times, high speed, a good rate of climb and high speed at low level rather than extended loiter, and long range or the carriage of a great deal of external fuel. The internal fuel fraction is, at 0.27, a trifle on the low side, while the RM-8 engine is not noted for being particularly economical. Nevertheless, the fighter configuration radius of action is given as 270nm (500km), although without qualification it is difficult to comment on this figure. Time to altitude is good — 1.67 minutes to 32,800ft (10,000m) — while wing loading is moderate, but thrust loadings are very ordinary by modern standards although they were par for the course at the time of design.

Where the Viggen really scores is in its reduced runway reliance. Sweden has taken a less optimistic view than most of the potential availability of fixed airfields, and deploys to off-base facilities, using suitable widened stretches of road, as a matter of routine. Short field performance was one of the main drivers at the design stage, and the canards, fixed surfaces with trailing-edge elevators, were adopted to overcome the worst faults of the delta wing. The result is excellent low-speed performance and short take off and landing distances. Landing is carried out as if on a carrier — straight in with no flare and a high vertical sink rate — with instruments used to obtain a precise ▶

Below: The JA 37 Viggen has fixed canard surfaces and is designed for operations from dispersed off-base facilities.

▶ touchdown point and reduce the scatter normal to land-based operations, and the landing roll is reduced by using reverse thrust. The stand-alone qualities of the Viggen when used in off-base exercises are very good.

The Viggen is often written up as some sort of superfighter, though its ancestry dates back a long way, possibly because it was the first of the canard fighters to attain large-scale service. This configuration does not endow it with any magical qualities, however: sustained turn rate at 20,000ft (6,100m) is a mere 6.3°/sec and while the fixed canards give good high AoA handling, being highly loaded they tend to stall quite quickly, and do little for the sustained turn rate. Instantaneous turn is believed to be quite good, though no firm figures have been released, but some years back the Viggen was evaluated

against the F-16 and came off very badly, as one might expect given the original design concepts coupled with a gap of several years between the two.

It has been suggested that the thrust reverser can be used to aid transient performance in flight. This seems most unlikely; a delta is pretty good at forcing overshoots without exotic measures of this nature, and the energy loss would be prohibitive in any case, though given a thrust: weight ratio well in excess of unity it might be a possibility. The thrust reverser has a drawback in combat, even without being used for reverse thrust, in that it exposes hot metal to the seekers of IR-homing missiles. The Viggen is pleasant to handle and was a first class fighter in its day, but that day has come and gone; it is still a formidable adversary, but not in the same league as the new generation,

although relaxed static stability, fly-by-wire and a new avionics fit might do wonders for it.

Armament
One 30mm Oerlikon KCA cannon with 150 rounds per gun, 2/4 AIM-9L Sidewinders, 2/4 Sky Flash

User
Sweden

Above: the Viggen is a good, although rather unexceptional fighter. Its unorthodox layout gives good high AoA handling and assists STOL capability.

Below: By current standards the Viggen has a modest turn capability, but the weapons fit of Skyflash and Sidewinder AAMs helps redress the balance.

Grumman F-14 Tomcat

Type: Two-seat fleet air defence interceptor and air superiority fighter. Some aircraft can be configured for reconnaissance.

The Grumman F-14 Tomcat can be described fairly as the first of the superfighters. More noted for its carriage of the ultra- long-range AIM-54 Phoenix air-to-air missile, it was d gned first and foremost as a Sparro -armed counter-air fighter for the   et air defence role; only later were   Phoenix missiles added. It is a remarkably fine close combat fighter, with virtually no angle of attack limitations, and despite its age and its size it has to be considered a contender for the title world's

greatest fighter.

A comparison with its predecessor, the F-4 Phantom, shows it to be bigger, considerably heavier and with a 1g performance envelope of almost exactly the same size, while thrust loadings are comparable and wing loadings much higher. There the similarity ends: the combination of variable-sweep wings and a large lifting surface, usually known as the pancake, at the rear of the fuselage, make a nonsense of orthodox wing loadings, the pancake alone providing some 40 per cent extra lifting area and combining with the lifting devices on the wings to produce a very manoeuvrable bird that even F-15 drivers do not care to slow down

Dimensions	F-14A	F-14D
Length (ft/m)	62.88/19.17	62.88/19.17
Span (ft/m)	64.13/19.55 max	64.13/19.55 max
Height (ft/m)	16.00/ 4.88	16.00/ 4.88
Wing area (sq ft/m²)	565/52.50	565/52.50
Aspect ratio	7.28 to 2.58	7.28 to 2.58
Weights		
Empty (lb/kg)	39,921/18,110	N/A
Takeoff (lb/kg)	58,571/26,570	N/A
Combat (lb/kg)	50,471/22,895	N/A
Power	2xTF30-414A	2xF110-400
Max (lb st/kN)	20,900/93.0	27,080/120.3
Mil (lb st/kN)	12,350/54.9	16,610/ 73.8
Fuel		
Internal (lb/kg)	16,200/7,350	16,200/7,350
External (lb/kg)	3,800/1,724	3,800/1,724
Fraction	0.28	N/A
Loadings		
Max thrust	0.71 — 0.83	N/A
Mil thrust	0.42 — 0.49	N/A
Wing takeoff (lb/sq ft/kg/m²)	104/506	N/A
Wing combat (lb/sq ft/kg/m²)	89/436	N/A
Performance		
Vmax hi	M = 2.31	M = 2.31
Vmax lo	M = 1.20	M = 1.20
Ceiling (ft/m)	56,000/17,100	N/A
Initial climb (ft/min/m/sec)	30,000/152	N/A
Takeoff roll (ft/m)	1,400/427	N/A
Landing roll (ft/m)	2,160/660	N/A
First flight	Dec 1970	N/A

and turn with.

The previous-generation Phantom is totally outclassed in a turning fight, as are many more modern fighters. The Tomcat can pull 7g at Mach 2 and still be holding this loading as the speed winds down through Mach 1. Corner velocity has not been released, but is somewhere below 300kt (556km/hr); nor have sustained turn rates been revealed, but all else being equal it is to be expected that very high g loadings can be sustained at the low end of the speed range, giving a very small turn radius and high turn rate. Equally, it may be assumed that the large size and great mass of the F-14 tend to reduce its transient performance. If the F-14A has a fault, it lies in the engine: the Pratt & Whitney TF30 needs nursing if it is to give of its best.

The F-14A has been in service virtually unchanged except for minor upgrades since 1972, but in 1988 the first F-14A Plus — essentially the F-14A with General Electric F110 engines — enters service. The new engines give a great deal more power, allowing a carrier takeoff to be made without afterburner, increasing the patrol time by a considerable margin and bringing the thrust loading at combat weight to over unity to make it without doubt one of the most formidable fighters in service anywhere. After 41 of the F-14A Plus have been delivered production will switch to the F-14D, with the advanced Hughes APG-71 radar and the latest digital avionics fit. Delivery of the F-14D will start in 1990.

The Tomcat is a weapon system in the true sense of the term. Its advanced AWG-9 radar can detect targets over 200nm (370km) away; it can track 24 targets, displaying 12 of of them on the screen in the rear cockpit, allocate threats and salvo up to six Phoenix missiles at six different targets in quick succession, providing semi-active homing for them ▶

Left: F-14A Tomcat of VF-84 Jolly Rogers is seen launching an AIM-54 Phoenix missile.

Below: Wings spread and flaps deployed, a Tomcat of VF-74 touches down on USS _Saratoga_.

▶ on a time-share basis. It can receive other contacts through the digital data link, and transmit data to other aircraft the same way. It works closely with E-2C Hawkeye AEW aircraft and the EA-6B Prowler electronic warfare aircraft, and can detect and destroy ultra-fast, ultra-high targets at 100nm (185km) ranges, and small sea-skimming missiles at 25nm (46km).

Without the long-range kill capability conferred by Phoenix, it is an incredibly good fighter: with Phoenix it is simply unique. As Soviet Foxbat pilot Viktor Belenko asked, 'how does one get near it?'

Despite its fantastic qualities, Tomcat's war record is nothing special, although this can be ascribed to poor pilot quality and lack of 100

Right: With burners blazing, an F-14 launches from *Saratoga*. The more powerful F-14D will be able to launch in dry power.

per cent serviceability when used in Iranian service against Iraq. It is believed to have been used in the AEW role, and no kill has been attributed to it in that war, though Iraqi fighters claim to have shot down at least three Tomcats. In US Navy service it has shown itself to be excellent in peacetime exercises, rarely allowing a threat aircraft to penetrate its defensive screen. It has flown TARPS reconnaissance missions, and armed patrols over Lebanon, and on the one occasion that it has fired in anger, on August 19, 1981, two Tomcats of VF-41 from USS *Nimitz* downed two Libyan Su-22 Fitters with Sidewinders in just 45 seconds.

Armament
20mm M61 Vulcan cannon with 675 rounds, maximum eight missiles in any combination from six AIM-54 Phoenix, six AIM-7F Sparrow and four AIM-9L Sidewinder; Sparrow and Sidewinder will be replaced by Amraam and Asraam.

Users
Iran, USA

Left: With wings fully swept, a tight formation of Tomcats follows a carrier wake.

93

Yakovlev Yak-38 Forger

Type: Single-seat carrier fighter. Two-seat trainer exists.

The advantages of vertical takeoff and landing have been apparent for many years, but are accompanied by certain inbuilt penalties, not the least of which is the difficulty of generating a thrust:weight ratio sufficient to get a worthwhile payload off the ground with enough fuel remaining to carry it a worthwhile distance. The American tail-sitters of the 1950s were followed by the Dassault Mirage IIIV, whose battery of eight lift engines became so much dead weight and wasted volume in cruising flight, even though this format did produce the

world's only VTOL fighter capable of Mach 2, but the only one that had worked had been the Hawker Siddeley Kestrel, later translated into the Harrier attack fighter, which used a single engine with four vectoring nozzles.

The Soviet Union was also working on a VTOL attack aircraft. First came the Yak-36 Freehand, no more than a technology demonstrator but one which appeared to fly quite satisfactorily, followed in due course by the Yak-38 Forger, known for many years as the Yak-36MP. The attack role had meanwhile been dropped, primarily due to ground erosion problems, it is believed, but the Soviet Navy had taken up the project.

Dimensions	Yak-38 Forger
Length (ft/m)	52.50/16.00
Span (ft/m)	24.78/ 7.49
Height (ft/m)	11.00/ 3.35
Wing area (sq ft/m²)	199/18.50
Aspect ratio	3.03

Weights	
Empty (lb/kg)	17,000/7,700
Takeoff (lb/kg)	24,000/10,900
Combat (lb/kg)	21,500/9,752

Power	1xAL-21 + 2 lift
Max (lb st/kN)	engines
Mil (lb st/kN)	17,985/80.0

Fuel	
Internal (lb/kg)	5,000/2,270
External (lb/kg)	N/A
Fraction	0.21

Loadings	
Max thrust	N/A
Mil thrust	0.75 − 0.84
Wing takeoff (lb/sq ft/kg/m²)	121/588
Wing combat (lb/sq ft/kg/m²)	108/527

Performance	
Vmax hi	M = 0.95
Vmax lo	M = 0.85
Ceiling (ft/m)	39,000/11,887
Initial climb (ft/min/m/sec)	15,000/76
Takeoff roll (ft/m)	V/STOL
Landing roll (ft/m)	V/STOL

First flight	circa 1971

Above: A line-up of Yak-38 Forger shipborne fighters spotted on the deck of the Soviet Navy's aircraft

The rapidly expanding Soviet Navy was lacking the essential facility of organic air power. It is not easy to build aircraft carriers, adapt aircraft for them and put to sea: naval aviation needs skill and experience, and aircraft with VTOL capability can operate from almost any ship that can accommodate more than a single helicopter, so it seems likely that Forger has been nothing more than a protracted experiment in the operation of fixed-wing aircraft from ships at sea and of VTOL aircraft in general.

This conclusion is the result of a process of elimination. Forger has neither the avionics nor the armament to be effective in the fleet air defence or air superiority roles; it has no integral gun, although pods can be carried on the inboard pylons, and the only missiles carried have been the AA-2 Atoll and AA-8 Aphid, both short-ranged IR-homers, while the radar is range only, with no search or track capability, and the nose shape could hardly accommodate a really capable radar and antenna.

The avionics deficiencies tend to rule out the reconnaissance role, and in the close air support mission it would be compromised by lack of payload — the four pylons are all on the non-folding section of the wing, which severely limits space. IR detection kit is carried, but its purpose is unknown. Thrust loading is not particularly good, and certainly not in the same league as that of the Harrier, while the wing loading is very high, and manoeuvrability can be assumed to be poor: In a straight contest, the British fighter could be expected to run out an easy winner.

Vertical flight capability is conferred by two lift engines, probably Koliesovs with around 8,000lb (3,630kg) of thrust each, and the propulsion engine is believed to be a L'yulka AL-21 as fitted in the Su-17/20 series, the smokiest Soviet engine in service; the latter has a vectoring nozzle. Forger is heavier than the Harrier, and its payload/radius performance cannot be anywhere near as good; it was thought at one time that it had no STOL capability, but that theory has since been disproved. It is also thought that Forger has no VIFF capability.

Forger has one gadget unknown in the West: called Eskem, its Russian acronym, it is an automatic ejection system which has to be switched on for take off and landing, processes which use a high-authority auto-stabilisation system. If certain limits are exceeded, Eskem promptly chucks the pilot overboard before the crash.

carrier *Kiev*. **Forger combines two lift engines with a vectored-thrust main engine for V/STOL capability.**

Armament
Two GSh-23 twin-barrel gun pods, two AA-2 Atoll or AA-8 Aphid heat-seeking missiles

User
USSR

Left: The hatch above the two lift engine intakes can be seen in the open position, just behind the cockpit.

McDonnell Douglas F-15 Eagle

Type: Single-seat air superiority fighter with ground-attack capability (A and C); two-seat trainers with full combat capability (B and D) optimised ground-attack (E) versions exist.

The F-15 Eagle owes much of its design to the MiG-25 Foxbat. It is now known that Foxbat was designed purely to intercept the B-70, but at one time it appeared that the Soviet Union had made a sudden advance across the board in technology, and produced a world beating tactical fighter which was being mass produced in enormous numbers. Foxbat was therefore to a large degree the driving force behind the F-15 Eagle and its slightly earlier naval companion, the F-14 Tomcat.

It was originally intended that the F-15 would outmatch the Foxbat right across the board, even to the Mach 3 top speed which Foxbat was thought to possess, and cost was no object. The result was the second of the American superfighters, and the second most costly fighter in history. There was some justification for the

Dimensions	F-15A	F-15C	F-15C with FAST pack
Length (ft/m)	63.75/19.43	63.75/19.43	63.75/19.43
Span (ft/m)	42.81/13.05	42.81/13.05	42.81/13.05
Height (ft/m)	18.46/ 5.63	18.46/ 5.63	18.46/ 5.63
Wing area (sq ft/m²)	608/56.50	608/56.50	608/56.50
Aspect ratio	3.01	3.01	3.01
Weights			
Empty (lb/kg)	28,000/12,700	29,180/13,240	30,300/13,700
Takeoff (lb/kg)	41,500/18,825	44,500/20,185	55,270/25,070
Combat (lb/kg)	35,680/16,184	37,772/17,135	43,668/19,810
Power	2xF100-100	2xF100-100	2xF100-100
Max (lb st/kN)	25,000/111.0	25,000/111.0	25,000/111.0
Mil (lb st/kN)	16,200/ 72.0	16,200/ 72.0	16,200/ 72.0
Fuel			
Internal (lb/kg)	11,635/5,280	13,455/6,103	23,205/10,526
External (lb/kg)	11,700/5,310	11,700/5,310	11,700/ 5,310
Fraction	0.28	0.30	0.42
Loadings			
Max thrust	1.20 — 1.40	1.12 — 1.32	0.90 — 1.15
Mil thrust	0.78 — 0.91	0.73 — 0.86	0.59 — 0.74
Wing takeoff (lb/sq ft/kg/m²)	68/333	73/357	91/444
Wing combat (lb/sq ft/kg/m²)	59/288	62/303	72/351
Performance			
Vmax hi	M = 2.5 +	M = 2.5 +	M = 2.5 +
Vmax lo	M = 1.2	M = 1.2	M = 1.2
Ceiling (ft/m)	65,000/330	65,000/330	65,000/330
Initial climb (ft/min/m/sec)	50,000/254	50,000/254	N/A
Takeoff roll (ft/m)	900/275	N/A	N/A
Landing roll (ft/m)	N/A	N/A	N/A
First flight	Jul 1972	N/A	N/A

expense lavished on the naval fighter; a carrier task force is possibly the highest-value single target on earth, and its airborne defensive assets are very limited. There was less excuse for the Eagle, except that the Soviets had to be seen to be beaten out of sight technologically. In 1976, when the secrets of the MiG-25 were laid bare, the US Air Force was left with what was arguably the world's best close combat fighter and no opponent that even came near it, and one so expensive that it was difficult to afford enough to go round.

The Eagle was given eight missiles, a gun, and a very advanced avionics fit. It could not match the F-14's capacity for simultaneous multiple destruction, but operated on a look-shoot-look sequence against a succession of targets, some of which could be beyond visual range. The wing loading was low and the thrust loading high — above unity throughout the fighter configuration weight range — while it was capable of out-matching many other aircraft using military power alone. At 15,000ft (4,600m) and Mach 0.9, it could hold a sustained turn rate of 11.8°/sec, increasing to 16.5°/sec at Mach 0.5; at Mach 0.9 and the same altitude its instantaneous turn rate reached 16.5°/sec. Design g loading was +9g and −3g. In the shape of the Streak Eagle, it broke many of the time-to-altitude records held by Foxbat, but the maximum speed at altitude had been reduced to Mach 2.5, somewhat slower than the original requirement.

The F-15 is an exhilarating aircraft to fly, though somewhat sensitive. Aerodynamically, its rate of roll is just a little on the slow side. It is equipped with HOTAS (hands on throttle and stick) controls which means that every control needed in a combat situation is to hand and can be located by touch, but needs a great deal of manual dexterity — the system is known by the pilots as the 'piccolo'. Like the F-14, it had trouble with the engine, which was rather touchy, and had to be nursed. It was ▶

Below: Giving a good view of its twin engines and twin fins McAir's two-seat demonstrator lifts off in minimum burner.

also, despite a good fuel fraction, a trifle short on endurance.

As with all fighters that undergo development, it has gained weight, with a corresponding loss of performance, although it is still far better than most. Perhaps the most controversial aspect of the F-15 arises from its capability, which was arrived at by giving it the best of everything at the design stage: it is big, a fact which has given rise to two conflicting schools of thought. The first is generally propounded by the pilots of F-16s, and goes, 'the Eagle can sneak into the fight unobserved until the moment comes when it is forced to turn. Then everyone within eight miles will see it, and point their noses and missiles at it'. The standard Eagle driver's rejoinder is, 'by the time we get to visual range, the opposition will have lost so many that they will be nothing but defensive'.

The combat record of the F-15 is very impressive. Starting on June 27, 1979, when four Syrian MiG-21s were downed by Israeli F-15s, the record at the time of writing stands at 58½ kills for no loss, of these 15½ victims have been MiG-21s, three

Above: streaming contrails from its wingtips, an F-15C Eagle reefs into a hard turn using military power only.

Below: F-15C Eagle with FAST packs attached to 6512th Test and Evaluation Squadron at Edwards AFB, California.

were MiG-23s, three were Foxbats (all downed with Sparrows), 34 not positively identified but believed to be mainly MiG-23s and Su-22s, and a solitary helicopter, all Syrian victims of the Israelis; and two Iranian F-4 Phantoms were shot down by Saudi Arabian F-15s. F-15s flew escort to the F-16s which mounted the raid on the Iraqi nuclear reactor at Osiriak, and were solely responsible for the long-range strike against the PLO headquarters in Tunis on October 1, 1985.

The F-15 is by now quite an elderly design, and is subject to a Multi-Stage Improvement Program, which naturally includes the avionics, but engines are also being upgraded the new F100- PW-220 is reported to give significant performance improvements, with faster acceleration and stall-free operation. The F-15C and -D were also fitted conformal fuel tanks to improve range and endurance, though at the expense of some performance, these FAST packs increasing the fraction to a huge 0.42.

In the pipeline is the F-15E, a two-seater optimised for the attack mission. Little data has been released on this variant to date except that the take-off gross weight is to be around 81,000lb (36,742kg). It must be doubted whether the low wing loading will permit the Eagle to fly effectively at the levels required to survive in a modern war scenario. The gust response is high, and one report has referred to a height of 500ft (150m) as an optimum, which is much too high for safe penetration.

Further in the future is the STOL Eagle, which is to combine canard foreplanes with vectored thrust nozzles to give a short — 1,500ft (457m) — runway requirement. It is anticipated that the STOL Eagle will have improved instantaneous manoeuvre capability, greater range and better acceleration. It is due to fly in 1988.

Armament
M61 Vulcan 20mm cannon with 675 rounds, four AIM-7F Sparrows and four AIM-9L Sidewinders (to be replaced by Amraam and Asraam)

Users
Israel, Japan, Saudi Arabia, USA

Israel Aircraft Industries Kfir

Type: Single-seat counter-air fighter with secondary attack capability. Two-seat trainer exists.

It can hardly be denied that the Kfir is a Mirage III variant which arose from the Israeli need to be self sufficient in fighter aircraft. The Mirage was available, known and trusted, but political considerations were preventing further deliveries. However, the Israelis managed to obtain engineering drawings for both the Atar engine and the Mirage 5 airframe. To manufacture both was a tall order,

and it was decided to concentrate on new airframes, to use the more powerful and more readily available General Electric J79, already in service in Israeli Phantoms.

Some original airframes stayed with the Atar; these were called Nesher, and about 40 are believed to have taken part in the October War in 1973. Most of the survivors were later renamed Dagger and exported, notably to Argentina, for whom they performed creditably but lost heavily in the South Atlantic War in 1982.

The original Kfir C.1 was intended as an attack type with secondary

Dimensions	Kfir C.2	Kfir C.7
Length (ft/m)	53.98/16.45	53.98/16.45
Span (ft/m)	26.96/ 8.22	26.96/ 8.22
Height (ft/m)	14.94/ 4.55	14.94/ 4.55
Wing area (sq ft/m²)	375/34.81	375/34.81
Aspect ratio	1.94	1.94
Weights		
Empty (lb/kg)	17,250/ 7,825	17,480/ 7,930
Takeoff (lb/kg)	23,300/10,570	23,540/10,680
Combat (lb/kg)	20,470/ 9,285	20,701/ 9,390
Power	J79-17	J79-J1E
Max (lb st/kN)	17,900/79.6	18,750/83.4
Mil (lb st/kN)	11,918/53.0	11,890/52.8
Fuel		
Internal (lb/kg)	5,670/2,572	5,670/2,572
External (lb/kg)	6,780/3,075	8,217/3,727
Fraction	0.24	0.24
Loadings		
Max thrust	0.77 — 0.87	0.80 — 0.91
Mil thrust	0.51 — 0.58	0.51 — 0.57
Wing takeoff (lb/sq ft/kg/m²)	62/303	63/306
Wing combat (lb/sq ft/kg/m²)	55/266	55/269
Performance		
Vmax hi	M = 2.2	M = 2.0
Vmax lo	M = 1.14	M = 1.14
Ceiling (ft/m)	50,000/15,250	58,000/17,700
Initial climb (ft/min/m/sec)	40,000/203	40,000/203
Takeoff roll (ft/m)	4,750/1,450	N/A
Landing roll (ft/m)	N/A	N/A
First flight	1973	N/A

counter-air capability but priorities were soon reversed, and the main production type became the C.2, which featured canard foreplanes and nose strakes. The Kfir was heavier than the Mirage deltas, but the additional power of the American engine more than offset that disadvantage. Thrust loadings of the Kfir are better, and this more than makes up for the wing loadings being higher.

Sustained turn at 15,000ft (4,600m) is 8.4°/sec at Mach 0.85 for the C.2 and 9.6°/sec at Mach 0.68 for the C.7. Instantaneous turn at the same altitude is 15.8°/sec at Mach 0.73 for the C.2 and 18.9°/sec at Mach 0.68 for the C.7.

The Kfir arrived just too late to participate in the October War, and apart from a few skirmishes it was able to play little part over the next few years. Then, over the Beka'a Valley in Lebanon in 1982, the brunt of the fighting was borne by the newer and far more potent F-15s and F-16s, though the Kfir did manage to notch up a few kills.

In recent times the Kfir has found a new role: the US Navy and Marine Corps are now using leased Israeli C.2s as adversary aircraft with the designation F-21A. The first adversary squadron, VF-43, based at NAS Oceana, is flying a sortie rate of five per day. A USMC squadron is planned for the Pacific Coast units, based at MCAS Yuma, where there is a TACTS/ACMI range.

Brochure figures for range are really quite surprising considering the large wetted area and the fact that the J79 is hardly the most efficient turbojet around. High-altitude interception with a supersonic run-out and two minutes combat at Mach 1.5 is stated to be possible at 419nm (775km), which even with three external jugs seems rather a long way. In clean configuration, the figure reduces to 186nm (346km).

Armament
Two 30mm DEFA cannon with 150 rounds per gun, two Sidewinder, Shafrir, or Python IR-homing missiles

Users
Argentina (Dagger), Colombia, Israel, USA

Left: Canard foreplanes were adopted on the Kfir to offset the faults of the delta wing.

Below: The Kfir was really a Mirage recast to take the J79 engine. The small nose strakes are an aid to stability.

General Dynamics F-16 Fighting Falcon

Type: Single-seat air superiority and counter-air fighter with initially limited but gradually increasing adverse weather capability, used by some air forces primarily in the ground-attack role. Two-seater fully combat-capable trainer versions exist.

In any discussion of fighter manoeuvrability the odds are that the F-16 will be used as a yardstick. Even 12 years after its first flight it remains the fighter to beat, and while it can be beaten fairly easily in some regimes, it is very difficult to counter across the board.

The F-16 dates from a time of reaction to the spiralling cost and size of new fighters. There could be little doubt that technology, however advanced, could be ground down by superior numbers, and numbers also produced confusion, which further degraded technology. The answer sought was to meet the opposition with something approaching numerical parity coupled with a distinct technical edge. The requirement was for an austere fighter; something that could be easily built in affordable quantities but could still outfly the adversary aircraft (in this case usually

Dimensions	F-16A	F-16C	F-16C MSIP
Length (ft/m)	49.25/15.01	49.25/15.01	49.25/15.01
Span (ft/m)	31.00/ 9.45	31.00/ 9.45	31.00/ 9.45
Height (ft/m)	16.58/ 5.05	16.58/ 5.05	16.58/ 5.05
Wing area (sq ft/m²)	300/27.88	300/27.88	300/27.88
Aspect ratio	3.20	3.20	3.20
Weights			
Empty (lb/kg)	16,234/ 7,364	17,960/ 8,150	17,960/ 8,150
Takeoff (lb/kg)	23,810/10,800	26,536/12,040	26,536/12,040
Combat (lb/kg)	20,324/ 9,220	23,050/10,455	23,050/10,455
Power	F100-100	F100-100	F100-400 or F100-220
Max (lb st/kN)	23,904/106.3	23,904/106.3	28,000/124.5
Mil (lb st/kN)	14,780/ 65,7	14,780/ 65.7	17,000/ 75.6
Fuel			
Internal (lb/kg)	6,972/3,162	6,972/3,162	6,972/3,162
External (lb/kg)	6,760/3,066	6,760/3,066	6,760/3,066
Fraction	0.29	0.26	0.26
Loadings			
Max thrust	1.00 − 1.18	0.94 − 1.08	1.06 − 1.21
Mil thrust	0.62 − 0.73	0.57 − 0.65	0.64 − 0.74
Wing takeoff (lb/sq ft/kg/m²)	79/387	88/432	88/432
Wing combat (lb/sq ft/kg/m²)	68/331	77/375	77/375
Performance			
Vmax hi	M = 2.0	M = 2.0	M = 2.0
Vmax lo	M = 1.2	M = 1.2	M = 1.2
Ceiling (ft/m)	50,000/15,250	50,000/15,250	50,000/15,250
Initial climb (ft/min/m/sec)	50,000/254	50,000/254	50,000/254
Takeoff roll (ft/m)	1,750/533	1,750/533	1,750/533
Landing roll (ft/m)	2,650/808	2,650/808	2,650/808
First flight	Jan 1974	Jun 1984	Jul 1982

considered to be MiG-21s) by a wide margin. A competition was instituted to examine the art of the possible, with a flyoff between two contenders, GD's YF-16 and Northrop's YF-17. The competition, which had become a full-scale programme, was narrowly won by the YF-16.

The new fighter was small, contained basic avionics for clear air fighting, and was armed with a 20mm M61 Vulcan cannon and two Sidewinders. Visibility from the cockpit was exceptional through a large teardrop canopy with low sills, a feature which had been revived with the F-14 and F-15. A single Pratt & Whitney F100 turbofan gave ample power, and the aerodynamic design had been optimised for close combat manoeuvring, though it used nothing in the way of exotic materials such as composites. It did, however, use various hi-tec features such as relaxed static stability coupled with quadruplex fly-by-wire controls, variable camber wings and leading edge strakes to clean up the flow, all of which combined to give tremendous manoeuvrability.

The new fighter was stressed to +9g and −3g, and its ability to perform a sustained 9g turn was widely acclaimed, but precisely what was this worth? In numerical terms, not a lot, being attainable only in a relatively small segment of the envelope and at moderate speeds and low altitudes. The difference between a 9g turn and an 8g turn at 400kt (740km/hr) is little more than 200ft (60m) in terms of radius, and about 3°/sec in rate of turn. On the other hand, any fighter pilot will tell you that this could be the difference between life and death in a defensive situation, though it is of less value when attacking: at 9g the pilot's faculties will be impaired, aiming is difficult and it is hard to hose the nose about at high angles of attack. ▶

Below: A two-seat F-16D demonstrates that it is fully combat capable as it pitches up to a high angle of attack.

▶ To counteract the effects of high loadings on the pilot, a radical solution was adopted. The seat was raked sharply back, the heel line raised, and the standard control column was replaced by a sidestick controller on the right-hand console. Pilot tolerance to high g is increased, but there are disadvantages: the raised heel line reduces the usable dash space, the sidestick cuts a good deal of console area, the seat rake has led to strained necks and shoulders, and in the event of a pilot becoming incapacitated in the right arm he cannot swap hands to retain control.

The F-16A and its two-seat counterpart, the F-16B, entered service and were duly hailed as a reversal of the trend toward increased size and cost. In USAF service this was true, as it formed part of what was known as the hi-lo mix with the F-15. The first overseas sales were to NATO members Belgium, Denmark, Holland and Norway, and while their pilots were delighted, the F-16 was immediately criticised for lack of all-weather capability, a critical consideration when one considers European weather.

The first consequence was that the F-16 was switched to the attack role as a primary function, the case with all Holland's F-16s. It was not difficult, the F-16 having been christened the 'swing fighter' for its ability to switch from one role to the other. The second was perhaps inevitable: a start was made on upgrading the avionics to cope, and the austere fighter, which had been fairly cheap, started to become more expensive. Trials have concentrated on BVR weapons, and many F-16s will carry Amraam when it enters service. The F-16 can now be considered the top end of the small fighter market.

A deliberate attempt was made to produce an even cheaper version, the

F-16/79, powered with the J79 turbojet, but this failed because no country was willing to accept such a blatantly downgraded version: if they wanted the F-16, they wanted the best version available.

Inevitably, a new variant appeared. The F-16C has improved avionics, including the much more capable APG-68 radar, which has more modes and better performance than the previous APG-66. And the current three-phase Multi-Stage Improvement Program (MSIP) involves an even more comprehensive avionics fit, as well as the more powerful F100-PW-220 or F110-GE-400 engine, which will improve performance all round.

Ironically, there has been a move toward a more austere F-16 in recent times, variously described as the F-16E, optimised for ground attack, or F-16CM, which will have the original APG-66 radar. The US Navy has ordered a small quantity of this version as its new adversary aircraft under the designation F-16N, which is to have the F110 turbofan, the APG-66 radar, no gun and some lower wing fittings of titanium instead of aluminium.

There is one other variant, the F-16XL or F-16F, a tailless delta with a high level of commonality with the basic aircraft which was developed to explore supersonic cruise and manoeuvre regimes. Heavier and with a much greater wing area consequent lower wing loading, but reduced thrust loading, it carries a colossal amount of fuel, the fuel fraction being 0.38. No order has been placed for the F-16F, but it may be developed in the ground attack role at some later date.

So far, only the Israelis have used the F-16 in combat. Israeli F-16s carried out the precision attack on the Iraqi nuclear reactor near Bagdhad in June 1981, and also featured in the Beka'a Valley action, where they are believed to have scored about 30 victories for no loss. Whatever its faults may be, it is the fighter that pilots itch to get their hands on.

Below: The F-16F is a tailless cranked delta developed to explore the supersonic cruise and manoeuvre regimes.

Armament
20mm M61 Vulcan cannon with 500 rounds, two or four AIM-9L Sidewinders, Shafrirs or Pythons (Amraams and Asraams in the future)

Users
Belgium, Denmark, Egypt, Greece, Holland, Indonesia, Israel, South Korea, Norway, Pakistan, Thailand, USA, Venezuela

Left: Israeli defences old and new: an F-16 of the Heyl Ha'Avir overflies the ancient Zealot stronghold of Masada.

McDonnell Douglas F/A-18 Hornet

Type: Single-seat multi-role carrier-borne and land-based fighter. Dedicated reconnaissance version under development.

If the F-16 became the aircraft to beat in close combat, the F/A-18 Hornet set new standards for multi-role fighters. It originated in the USAF lightweight fighter competition won by the F-16 in 1974, the Hornet having been developed from the Northrop YF-17 that came a very close second. At that time the US Navy was in the market for a new fighter/attack aircraft to replace its elderly Phantoms and Corsairs, and it was widely thought that the winner of the competition would be selected.

Instead, the Navy took a long and hard look at both contenders before selecting the YF-17 as having more development potential and meeting its requirements more closely. Northrop teamed with McDonnell Douglas, who had vast experience of building carrier fighters, and the F/A-18 was developed as a joint effort. At the same time, Northrop were to develop the land-based F-18L for the export market.

Dimensions	F/A-18A
Length (ft/m)	56.00/17.07
Span (ft/m)	37.50/11.43
Height (ft/m)	15.29/ 4.66
Wing area (sq ft/m²)	400/37.17
Aspect ratio	3.52

Weights	
Empty (lb/kg)	21,830/ 9,900
Takeoff (lb/kg)	35,800/16,240
Combat (lb/kg)	30,370/13,775

Power	2xF404-400
Max (lb st/kN)	16,000/71.2
Mil (lb st/kN)	10,600/47.2

Fuel	
Internal (lb/kg)	10,860/4,925
External (lb/kg)	7,000/3,175
Fraction	0.30

Loadings	
Max thrust	0.89 − 1.05
Mil thrust	0.59 − 0.70
Wing takeoff (lb/sq ft/kg/m²)	90/437
Wing combat (lb/sq ft/kg/m²)	76/371

Performance	
Vmax hi	M = 1.8
Vmax lo	M = 1.01
Ceiling (ft/m)	50,000/15,250
Initial climb (ft/min/m/sec)	50,000/254
Takeoff roll (ft/m)	N/A
Landing roll (ft/m)	N/A

First flight	Jun 1974

It was intended to produce two distinct types, optimised for fighter or attack work, the main difference being the avionics and cockpit layout, but McDonnell Douglas made that unnecessary: using their cockpit experience with the F-15, they combined HOTAS with CRT displays to produce what was in effect a new-generation cockpit with hardly any old fashioned dials and tape instruments; instead, there were three screens on which selected information could be called up at need. The aircraft was also designed to carry two Sparrows conformally for the fighter mission, and to be quickly changed for the attack mission by substituting FLIR and laser designator pods in their places, while the Hughes APG-65 multi-mode radar has numerous high-quality air-to-air and air-to-ground modes.

The 1974 air combat fighter competition continues in terms of interservice rivalry, and the merits of the Hornet against the Fighting Falcon are still a matter for debate. Initially, at any rate, the Hornet's radar and weapons fit was superior, with two Sparrows and two Sidewinders as standard, while the F-16 had no answer to the BVR weapon and its radar was comparatively short on both range and capability. This has since been rectified, but the Sparrow is a big missile, and is carried with less penalty by the larger Hornet than the smaller Fighting Falcon, to which conformal carriage is denied.

At close range the odds shorten. In purely numerical terms the F-16 appears to be the better of the two, but the outside of the envelope is seldom if ever reached in combat. The F-18 cannot match the sustained turning capability of the F-16, but has better acceleration between Mach 0.8 and Mach 1.2, is at least as good in the roll ▶

Left: An F/A-18A Hornet of VMFA-314 Black Knights, based at El Toro, fires a Sidewinder.

Below: The remarkably sleek lines of the Hornet are seen to advantage in this study of a VFA-113 Stingers formation.

and is distinctly snappier in pitch. It has virtually no AoA limitations, while the F-16 is believed to be a bit short on high-AoA capability. The engines on the F-18 were much more resistant to hard use than that on the F-16, but the new generation of F-16 engines will level the score on this point. In BVR combat, the Hornet's superior avionics should give it the

Left: A Canadian CF-18 shows the typical Hornet armament of two Sparrows and two 'winders. The dummy canopy is painted on for aspect deception.

edge, while at close range the fight will go to the best pilot.

In the export field the F/A-18 was often directly opposed by the F-16, and in most cases the F-16 was selected, but it should be remembered that the F-16 was considerably cheaper, and it is very noticeable that whenever the operational requirements were particularly stringent, the F/A-18 was selected — so far by Australia, Canada, and Spain.

The Hornet has not seen action, though it is credited with the destruction of a TA-4 Skyhawk which was unfortunate enough to get in the way of a jettisoned bomb rack. Its great merits are that it can look after itself en route to the target, being quite a formidable fighter even with a load on, while on the return trip, where a dedicated attack aircraft might need a fighter escort, a Hornet swarm becomes a fighter sweep in its own right. It seems a pity that the lighter but equally powerful Northrop F-18L found no buyers, as it would have been an even more potent fighter than its navalised and more heavily loaded cousin.

The two-seat F/A-18B is fully combat capable, but carries 600lb (272kg) less fuel and is consequently a bit shorter on range and endurance. A dedicated reconnaissance variant is under development, while the F/A-18C and D will shortly make their appearance, the change in designation indicating that they are compatible with Amraam, ASPJ, and IIR Maverick. A two-seat night and adverse weather variant has been considered to replace the ageing A-6 Intruder, with an advanced avionics fit dedicated to ground attack, but it seems unlikely that this will proceed.

Armament
20mm M.16. Vulcan cannon with 570 rounds, two AIM-7M Sparrow and two AIM-9L Sidewinder (Amraam and Asraam later)

Users
Australia, Canada, Spain, USA

Left: The leading edge strake shows up well from this angle as a VMFA-323 Death Rattlers F-18 launches from *Coral Sea*.

Panavia Tornado ADV

Type: Two-seat long-range interceptor. Interdiction/strike version in service, reconnaissance and electronic warfare versions proposed.

The air-defence Tornado was developed from the IDS deep-penetration interdiction and strike bomber Tornado, and while many reasonable attack aircraft have been developed from fighters, the reverse is very rare. The reason in this case is the need to defend both the United Kingdom and the airspace around it out to a considerable distance: the priorities were long range and endurance, with extended periods us-

ing afterburner if necessary; plenty of missiles to give combat persistence; and an advanced radar and avionics system to permit multiple target engagement in any weather, beyond visual range, and in the face of intense ECM. These qualities are similar to those required by a carrier fighter, and in fact the Grumman F-14 was evaluated for the role, but it was decided that Tornado was more cost-effective and had a far more modern avionics suite.

In terms of cold figures, Tornado F.3 is nothing very special. Wing loadings are high, thrust loadings are low and the fuel fraction is moderate; it appears to be no more than a vehi-

Dimensions	Tornado F.3
Length (ft/m)	59.25/18.06
Span (ft/m)	45.58/13.89 max
	28.21/ 8.60 min
Height (ft/m)	18.31/ 5.53
Wing area (sq ft/m²)	323/30.01
Aspect ratio	6.43 to 2.46

Weights	
Empty (lb/kg)	31,500/14,290
Takeoff (lb/kg)	50,200/22,770
Combat (lb/kg)	43,950/19,935

Power	2xRB199 Mk104
Max (lb st/kN)	16,920/75.2
Mil (lb st/kN)	9,656/42.9

Fuel	
Internal (lb/kg)	12,500/5,670
External (lb/kg)	14,391/6,530
Fraction	0.25

Loadings	
Max thrust	0.67 − 0.77
Mil thrust	0.38 − 0.44
Wing takeoff (lb/sq ft/kg/m²)	155/759
Wing combat (lb/sq ft/kg/m²)	136/664

Performance	
Vmax hi	M = 2.27
Vmax lo	M = 1.20
Ceiling (ft/m)	50,000/15,250 +
Initial climb (ft/min/m/sec)	40,000/203 +
Takeoff roll (ft/m)	2,500/760
Landing roll (ft/m)	1,200/370

First flight	Oct 1979

cle to take missiles to a suitable launch position. But the figures are misleading: no-one is going to pretend that a Tornado will outfight an F-16 in the close combat arena, but it would not be outclassed as completely as the bare figures suggest. The variable geometry wing is a great advantage, having more high-lift devices than any other Mach 2 capable fighter, and they combine to give it what is probably the highest lift coefficient of any fast jet. The result is that it out-manoeuvres any other aircraft in RAF service, and is predicted to become a very effective performer in the close combat arena. The F.2 had only four wing settings, but the F.3, like the Tomcat, has fully variable automatic sweep, which keeps the wings at the optimum angle. In addition, a fair amount of body lift is created by the broad fuselage.

Nor are the low thrust loadings and fuel fraction particularly relevant in the case of Tornado. The engines were designed for economy, in-flight refuelling is available, and it is noticeable that the maximum external fuel load exceeds that carried internally. And while the engines are not the optimum for an air superiority fighter, acceleration is remarkably good, Tornado F.3 out-accelerating both the Lightning and the Phantom quite comfortably. This is partly a ▶

Below: Tornado F.2 displays the assortment of high lift devices which give it such a remarkable combat performance.

result of its aerodynamic cleanness and partly due to a good fineness ratio.

The avionics are superb, and the back-seater can be given the broad tactical picture via digital data link. He is responsible for structuring the combat, and at close quarters becomes a second pair of eyes looking out. The combination of high lift wings and good acceleration makes Tornado a surprisingly good performer in the close combat arena, and it is anticipated that when the day comes, the Tornado knockers are due for a big shock.

Only 18 Tornado F.2s were built, all serving with No 229 OCU at RAF Coningsby. The remaining fighters are F.3s, which are also to enter Saudi Arabian and Omani service.

Armament
27mm Mauser cannon, four Sky Flash or Sparrow, four Sidewinders (Amraam and Asraam in future)

Users
Oman, Saudi Arabia, UK

Below: Tornado F.2 in battle gear, with four Skyflash and two Sidewinders. Tornado F.3 carries two more Sidewinders.

British Aerospace Hawk

Type: Two-seat trainer equipped for clear weather air defence (Hawk T.1A) and single-seat light air defence fighter and barrier patrol aircraft (Hawk 200).

It is common for military trainers to be used as light attack aircraft, but it is very unusual to equip a trainer for the air to air role. The Hawk, though, is a quite exceptional aeroplane. It has a quite ordinary fuel fraction, coupled with an unaugmented Adour turbofan; it is small; and it carries little in the way of avionics. Thrust loading is low, and wing loading low to moderate, although these are coupled with a high aspect ratio and a high co-efficient of lift. It has by modern fighter standards a poor rate of climb, and, except in a dive, is firmly subsonic.

At the same time, it can carry a 30mm gun pod on the centreline and a Sidewinder under each wing; it is, as would be expected from a trainer, very manoeuvrable, and can turn very well, sustaining 14°/sec at 400kt (740km/hr) in full fighter configuration; it has been designed for +9g and −3g, and has been cleared to +8g with full internal fuel and weapons load; the economical Adour turbofan gives it a remarkable range and endurance; and, finally, it is a small radar and infra-red target as well as being visually small.

It is immediately obvious that the ▶

Dimensions	Hawk T.1A	Hawk series 60	Hawk 200
Length (ft/m)	38.92/11.86	38.92/11.86	37.33/11.38
Span (ft/m)	30.83/ 9.40	30.83/ 9.40	30.83/ 9.40
Height (ft/m)	13.16/ 4.00	13.16/ 4.00	13.67/ 4.17
Wing area (sq ft/m²)	180/16.69	180/16.69	180/16.69
Aspect ratio	5.28	5.28	5.28
Weights			
Empty (lb/kg)	7,450/ 3,380	8,015/3,635	8,765/3,975
Takeoff (lb/kg)	11,300/ 5,125	11,650/5,285	12,630/5,730
Combat (lb/kg)	9,875/ 4,480	10,187/4,620	11,163/5,065
Power	Adour 151	Adour 861	Adour 871
Max (lb st/kN)	N/A	N/A	N/A
Mil (lb st/kN)	5,300/23.6	5,700/25.3	5,845/26.0
Fuel			
Internal (lb/kg)	2,849/1,292	2,927/1,330	2,927/1,330
External (lb/kg)	1,560/708	2,966/1,345	2,966/1,345
Fraction	0.25	0.25	0.24
Loadings			
Max thrust	N/A	N/A	N/A
Mil thrust	0.47 − 0.54	0.49 − 0.56	0.46 − 0.52
Wing takeoff (lb/sq ft/kg/m²)	63/307	65/317	70/343
Wing combat (lb/sq ft/kg/m²)	55/268	57/277	62/303
Performance			
Vmax hi	M = 0.98	M = 0.98	M = 0.98
Vmax lo	M = 0.94	M = 0.94	M = 0.94
Ceiling (ft/m)	50,000/15,250	50,000/15,250	50,000/15,250
Initial climb (ft/min/m/sec)	6,000/30.5	9,300/47.0	12,000/61.0
Takeoff roll (ft/m)	2,000/610	1,750/520	1,500/457
Landing roll (ft/m)	2,000/610	1,900/580	1,900/580
First flight	Aug 1972	N/A	May 1986

▶ Hawk might make an adequate air defence fighter in the service of a Third World country opposed by a low-grade threat, but less apparent how it might fit into the RAF's force structure. However, as we have seen, confusion degrades technology, and numbers contribute to confusion. An enemy attack on the British Isles can only be effectively delivered from long range and at low altitude. Laden attack aircraft are subsonic, and are therefore well within the performance envelope of the Sidewinder-armed Hawk.

The Hawk will in fact be used as a second line of defence, in one of two modes. The first is the combat air patrol in clear weather, aided by GCI or AEW, a task for which its long endurance fits it well; the second is the Mixed Fighter Force Concept (MFFC), which has been widely tested in recent years. MFFC originated in the early 1970s during air defence exercises, when Phantoms at high altitude used their radars in look-down mode to detect low flying intruders, after which they would direct other Phantoms onto them. MFFC involves attaching two or three Hawks to a Phantom which will operate in the director role for their interceptions. The small and agile Hawks coupled with the BVR kill capability of the Phantom should make a deadly combination.

The basic Hawk in RAF service is the T.1; modified to carry Sidewinders it becomes the T.1A, and training unit aircraft have been allocated shadow squadron numbers 63, 79, 151 and 234. In time of war they would be flown by instructors. The Series 50 Hawk, an export version aimed at the light attack/trainer requirement with improved avionics and a more powerful Adour, was followed by the Series 60, with more thrust, better handling and other improvements, while the Hawk 100 is dedicated to the attack role.

The latest Hawk variant is the single seat Hawk 200 which, while advertised as a fighter, can be optimised for almost any role by using a variable front fuselage section. Hawk 200 made its first flight from Dunsfold on May 19, 1986, but just 28 flights later it crashed on July 2 of the same year. Hawk 200 will carry a radar with a 24in (610mm) antenna, either the Ferranti Blue Fox or Blue Falcon or the Emerson APG-69. It will carry two guns internally — the 27mm Mauser cannon used by Tornado, the standard 30mm Aden, or the 25mm Aden as carried by Harrier GR.5 — and it has the high lift Phase III combat wing, which adds to its already impressive manoeuvrability, while at low level it is reported to be able to sustain 8g 'forever'. The Phase III wing has a small leading edge droop, and an enlarged leading edge radius, with improved flaps.

One further Hawk variant is the BAe/McDonnell Douglas T-45 Goshawk, which will be the US Navy's basic trainer for many years to come.

Armament
One or two cannon (see text), two or four AIM-9L Sidewinders

Users
Abu Dhabi, Bahrain, Dubai, Finland, Indonesia, Kenya, Kuwait, Saudi Arabia, UK, USA, Zimbabwe

Above: RAF Hawk trainers have been equipped with Sidewinder missiles and gun pods housing a 30mm Aden cannon for the secondary intercept role.

Below: With the new designation Hawk T.1A, the Sidewinder-armed Hawks will operate in conjunction with radar-equipped Phantoms to intercept subsonic attackers.

Shenyang F-8 and F-8 II Finback

Type: Single-seat twin engined multi-role fighter capable of all-weather interception and air superiority, with secondary interdiction and close air support capabilities.

The original J-8 Finback bore a strong resemblance to the Mikoyan Ye-152 Flipper, first seen in public in 1961, which was hailed by certain sections of the Western Press as the first Soviet Mach 2 fighter big enough to have a reasonable payload/range; however, it could not carry a big enough radar antenna and was dropped in favour of the Su-15. Whether there is a direct link between Finback and Flipper is not known; in any case, the first photograph of the F-8 I was only released to the West in 1984. Apparently a scaled-up version of the F-7, it featured a pitot nose intake.

First indication of the F-8 II came from a defector in the early 1980s who stated that Finback was undergoing a nose job similar to that carried out on the F-6 to turn it into the A-5. This report was followed by an announcement from the Xinhua News Agency that a J-8 with side intakes had flown in May 1984. This new configuration gave a performance im-

Dimensions	F-8 II Finback
Length (ft/m)	50.86/15.5
Span (ft/m)	34.45/10.50
Height (ft/m)	17.23/ 5.25
Wing area (sq ft/m²)	358/33.3
Aspect ratio	3.32

Weights	
Empty (lb/kg)	18,000/ 8,165
Takeoff (lb/kg)	31,000/14,061
Combat (lb/kg)	26,625/12,077

Power	
	2xR-33D
Max (lb st/kN)	18,300/81.3
Mil (lb st/kN)	11,250/50.0

Fuel	
Internal (lb/kg)	8,750/3,969
External (lb/kg)	2,000/907
Fraction	0.28

Loadings	
Max thrust	1.18 − 0.84
Mil thrust	0.73 − 0.84
Wing takeoff (lb/sq ft/kg/m²)	87/422
Wing combat (lb/sq ft/kg/m²)	74/363

Performance	
Vmax hi	M = 2.3
Vmax lo	M = 1.2
Ceiling (ft/m)	55,000/16,750
Initial climb (ft/min/m/sec)	50,000/254
Takeoff roll (ft/m)	1,640/500
Landing roll (ft/m)	1,475/450

First flight	1977

provement and also allowed a modern AI radar and more powerful engines to be fitted. Finally, a model with a large, folding ventral fin was shown at Farnborough 86, and brochures issued there form the basis for this account.

The tabular data assumes the use of the WP-7BM turbojet, while the internal fuel capacity has been calculated from the given data. It is entirely possible that the engines are more powerful and that much more internal fuel is carried, in which case the takeoff and combat weights would be greater. Maximum range is stated as 1,187nm (2,200km), but this is unqualified, as are the takeoff and landing speeds of 175kt (325km/hr) and 156kt (290km/hr) respectively.

Above and below: The F-8 II is the latest Chinese fighter to be revealed to the West. The ventral fin is folded sideways for takeoff.

Armament
One 23mm twin-barrel cannon with 200 rounds; IR and SARH missiles.

User
China.

Mikoyan MiG-31 Foxhound

Type: Two-seat air defence interceptor with all-weather capability.

One thing that can be said with total certainty about the MiG-31 Foxhound is that it has been developed from the MiG-25 Foxbat. Virtually everything else is speculation, stemming either from the various Western intelligence agencies or from other experts in the field of military aviation, and given the past record of such sources we can only accept the published data until such time as it can definitely be proven or refuted. It is known that the Foxbat was developed as a counter to the B-70 Valkyrie, and it has been postulated that the Foxhound was the obvious

follow up to counter the Rockwell B-1A which, with its mission profile combining Mach 2.2 at high level with high subsonic speeds at low level, would have been — and still would be —a difficult interception problem.

What is often overlooked is that the B.52 switched from high to low level penetration at about the time that Valkyrie was cancelled, and even that lumbering giant, firmly subsonic and with an enormous radar cross-section, and whose idea of low level was 1,500ft (460m) above ground level, would still have been an awfully difficult target for an air force almost totally dependent on ground control, especially in view of the increasing ingenuity of its countermeasures. It is

Dimensions	MiG-31 Foxhound
Length (ft/m)	72.50/22.10
Span (ft/m)	46.00/14.02
Height (ft/m)	18.50/ 5.64
Wing area (sq ft/m²)	662/61.52
Aspect ratio	3.20

Weights	
Empty (lb/kg)	47,500/21,545
Takeoff (lb/kg)	85,000/38,556
Combat (lb/kg)	69,000/31,300

Power	2xRD-F
Max (lb st/kN)	32,000/142.5
Mil (lb st/kN)	22,000/ 98.0

Fuel	
Internal (lb/kg)	32,005/14,520
External (lb/kg)	6,868/ 3.120
Fraction	0.38

Loadings	
Max thrust	0.75 — 0.93
Mil thrust	0.52 — 0.64
Wing takeoff (lb/sq ft/kg/m²)	128/627
Wing combat (lb/sq ft/kg/m²)	104/509

Performance	
Vmax hi	M = 2.40
Vmax lo	M = 0.95
Ceiling (ft/m)	75,000/22,900
Initial climb (ft/min/m/sec)	41,000/208
Takeoff roll (ft/m)	N/A
Landing roll (ft/m)	N/A

First flight	1975

Below: Designed to counter low-flying interdictors, the MiG-31 clearly shows its derivation from the earlier MiG-25.

much more likely that Foxhound was developed to counter low-flying interdictors in general rather than the B-1 in particular, and it is open to question whether Foxhound is an effective answer to the B-1B.

To move on from Foxbat, the first need was look-down radar capability, either in the fighter or in the form of AEW. Then the worst failings of Foxbat — high fuel consumption, severe structural limitations, and the almost total dependence on GCI via data link — had to be fixed.

High fuel consumption could be cured by using more efficient engines, but probably at the expense of the very high speed capability. The substitution of titanium for nickel steel would save weight and alleviate the structural limitations to some degree, although not enough to permit Foxhound to indulge in close combat. Finally, a Doppler look-down radar with its own operator, coupled with the long-distance detection capability of the 1I-76 Mainstay AEW&C aircraft would allow MiG-31 crews a measure of autonomy in their tactics.

The MiG-31's radar is believed to have been derived from Western technology, and has a search range of 165nm (305km) and a tracking range of 145nm (270km), but the radar cross-section of the target is not given; it is also believed to have a multiple target engagement facility, and certainly Foxhound has IR detection kit, presumably for use against the B-1B. Its main missile armament is the large AA-9, which has the remarkably short range of 25nm (46km) at high and half that at low altitude from head-on. DoD sources ascribe a Vmax interception radius of 400nm (740km) to Foxhound, and a patrol radius of double that.

The MiG-31 will be standard equipment for the component of the reorganised and upgraded interceptor force that is dedicated to strategic air defence.

Armament
Four AA-9 missiles, possibly with four smaller AA-8 missiles

User
USSR

Mikoyan MiG-29 Fulcrum

Type: Single-seat all-weather counter-air fighter with secondary attack capability. Fully combat capable two-seat trainer being built.

Since the appearance of the first satellite pictures of Fulcrum at the Ramenskoye test centre appeared several years ago, the MiG-29 has been the subject of intense speculation by the West. The speculation did not end when six of these fighters visited the Finnish Air Force base at Kuopio-Russala in July 1986: excellent photographs of the type became available for the first time, but little firm information was forthcoming, apart from the fact that a

more exact estimate of the overall dimensions became possible. The accompying data table is still fairly speculative, therefore, and the figures cannot be regarded as more than approximations based on what can be observed and what can be inferred.

It was at first thought that the MiG-29 was to be a Soviet equivalent of the F-16, albeit twin-engined, but it now appears that it is very close in dimensions and probably fairly close in weight to the bigger F/A-18 Hornet. Intelligence sources deemed it to be an uncompromised air superiority fighter, with a look-down/shoot-down weapon system based on APG-65. It was thought to

Dimensions	MiG-31 Fulcrum
Length (ft/m)	56.43/17.20
Span (ft/m)	37.73/11.50
Height (ft/m)	14.44/ 4.40
Wing area (sq ft/m²)	400/31.76
Aspect ratio	3.56

Weights	
Empty (lb/kg)	22,500/10,205
Takeoff (lb/kg)	35,000/15,875
Combat (lb/kg)	30,125/13,665

Power	
	2xR-33D
Max (lb st/kN)	18,300/81.3
Mil (lb st/kN)	11,250/50.0

Fuel	
Internal (lb/kg)	9,750/4,425
External (lb/kg)	2,000/ 907
Fraction	0.28

Loadings	
Max thrust	1.05 — 1.21
Mil thrust	0.64 — 0.75
Wing takeoff (lb/sq ft/kg/m²)	88/427
Wing combat (lb/sq ft/kg/m²)	75/368

Performance	
Vmax hi	M = 2.2
Vmax lo	M = 1.06
Ceiling (ft/m)	55,000/16,750
Initial climb (ft/min/m/sec)	50,000/254
Takeoff roll (ft/m)	N/A
Landing roll (ft/m)	N/A

First flight	1977

carry two cannon of either 23mm or 30mm in the wing roots, plus six missiles — either two Apex and four Aphid or six of the new AA-10.

It has since become obvious that only one cannon is carried, in the port wing root, and that four missiles may be the norm. Sustained turn has been assessed as 16°/sec at Mach 0.9 and 15,000ft (4,600m), and instantaneous turn 21°/sec at the same height and speed. Various estimates of operational radius have appeared, but it seems unlikely that it is as good as the Hornet.

Overall size and similar configuration — twin canted fins, comparable aspect ratio, and large extensions to the wing leading edge root — make comparisons with the Hornet inevitable, but the Western design that the MiG-29 most resembles is the fixed-wing F-14 variant that was proposed at one design stage; the wing shape is similar, and the effect is enhanced by the twin engine tunnels. The wide spacing of the engines, while good from a survival viewpoint, adds to the wetted area and consequently to drag, although it does give extra keel area, but hardly allows room for stores to be carried without an unacceptable degree of interference drag.

Left: MiG-29s airborne, showing their widely separated engines and sharply raked intakes.

Below: One of the quartet of Fulcrums that paid a surprise visit to Finland in July 1986.

It is tempting to speculate that the Soviet engine designers have cut it a bit too close, and that the engine needs an absolutely straight inlet for optimum functioning at high speeds. On the subject of inlets, however, the most fascinating feature is the doors that close when weight is on the nosewheel. It has been speculated that they can be closed in flight to prevent enemy radars picking up reflections from the compressors, but the obvious answer is that it is an anti-fod measure. The Soviets, being pragmatists, have probably adopted the device to prevent ingestion of snow and slush and to allow operation from hard surfaces of poor quality, and in time of war runways are likely to be littered with debris.

Fulcrum represents a new generation of Soviet fighters, and the view from the cockpit, while not up to the latest Western standards, represents a major advance over their previous aircraft. Whether CRT displays are installed remains to be seen; Fulcrum has been hailed as a true multi-role fighter, and if that is the case it would need modern multi-function displays to ease pilot workload. One feature that has aroused considerable comment is the presence of what appears to be an infra-red ball sensor on the nose, just ahead of the cockpit. While it has been used on many Western fighters, IR detection has not really proved a great success. It is suspected that the Soviet Union is ahead of the West in this field; it is equally possible that, with the B-1B entering service in increasing numbers, their need is greater.

Fulcrum entered service in 1984, and, unusually, it is to be exported in a full-up rather than a degraded version to both Syria and India. This is quite out of keeping with normal Soviet practice, but any attempt to explain it would just be more speculation.

Armament
One cannon, probably 23mm single-barrel, in the port wing root, four or possibly six AA-10 medium-range missiles, or two AA-7 Apex and four AA-8 Aphids

Users
India, Syria, USSR

Sukhoi Su-27 Flanker

Type: Single-seat all-weather air superiority fighter with some attack capability.

Developed simultaneously with Fulcrum, and bearing a strong family likeness despite the fact that to different design bureaux are involved, the Su-27 Flanker are also spotted by satellite at the Ramenskoye test site in 1977. The similarities were such that Western intelligence jumped to the conclusion that the two aircraft, code-named Ram-K (Flanker) and Ram-L (Fulcrum), comprised a hi-lo mix like the F-15 and F-16, with Fulcrum thought to be about the same size and capability as the F-16 — an assumption that has since been proved wrong: Fulcrum is more like the F/A-18 Hornet — while Flanker was likened to the F-15. Flanker has not yet been seen in the West, and details of it are even more conjectural than those given for Fulcrum, which has at least made one appearance.

According to data so far released, Flanker is slightly bigger than the F-15 and considerably heavier, wing loading and aspect ratio are higher, thrust loading is similar, ceiling is lower and initial climb rate is better. Turn capability is better — sustained turn at Mach 0.9 and 15,000ft is reported as 17°/sec and instantaneous turn as 23°/sec — and at least comparable with that of the up-engined F-16. Fulcrum is not quite so

Dimensions	Su-27 Flanker
Length (ft/m)	67.3/20.5
Span (ft/m)	47.50/14.48
Height (ft/m)	18.04/ 5.50
Wing area (sq ft/m²)	689/64
Aspect ratio	3.27

Weights	
Empty (lb/kg)	33,070/15,000
Takeoff (lb/kg)	49,600/22,500
Combat (lb/kg)	42,435/19,250

Power	2xR.29
Max (lb st/kN)	28,100/125
Mil (lb st/kN)	15,700/70

Fuel	
Internal (lb/kg)	14,330/6,500
External (lb/kg)	6,864/3,310
Fraction	0.29

Loadings	
Max thrust	1.13 − 1.32
Mil thrust	0.63 − 0.74
Wing takeoff (lb/sq ft/kg/m²)	72/352
Wing combat (lb/sq ft/kg/m²)	62/301

Performance	
Vmax hi	M = 2.30 +
Vmax lo	M = 1.10 +
Ceiling (ft/m)	60,000/18,300
Initial climb (ft/min/m/sec)	60,000/305
Takeoff roll (ft/m)	N/A
Landing roll (ft/m)	N/A

First flight	1977

good, despite the fact that the smaller aircraft might be expected to be superior, and Flanker's high aspect ratio appears to have been tailored for sustained turn. Acceleration is also good, reportedly some 20 per cent better than that of Flogger.

Flanker has a multi-mode radar which uses Doppler modes to give a true look-down/shoot-down capability and is believed to be used on Western technology. It has a track-while-scan facility, and performance is reported to be 130nm (240km) search range and around 100nm (185km) tracking range. The avionics also include a digitial data link to the ground detection and control systems and airborne early warning. Flanker is believed to carry IR search and tracking equipment, probably very similar to that seen on Fulcrum.

Flanker is replacing Fiddler and Flagon in Soviet service and is known to be operating from an airfield near the Baltic coast which has a runway marked out to simulate the flight deck of a 65,000-tonne aircraft carrier similar to that of the new aircraft carrier *Kremin*. The obvious inference is that a naval Flanker will be developed, but where Soviet aviation is concerned nothing is ever obvious, and using Flanker as a carrier fighter is going to be very difficult for a nation without non-VSTOL fast jet carrier experience.

Armament
One 23mm or 30mm cannon, four, six or eight AA-10 missiles, or four AA-10 and four AA-11 dogfight missiles.

User
USSR.

Below: Flanker is the newest Soviet air superiority fighter and has performance comparable to that of the F-15 Eagle. It became operational with air defence regiments early in 1986.

Below: The Soviet Union has been keeping the Su-27 Flanker under wraps; this picture, taken from a television screen, is the best currently available.

Dassault-Breguet Mirage 2000

Type: Single-seat interceptor/air super-iority fighter. Other variants optimised for the attack (2000C1), two-seat trainer (2000B), low-level and nuclear strike (2000N) and reconnaissance (2000R) roles.

The origins of the Mirage 2000 are confused, to say the least. After the variable geometry Mirage G, Dassault's next project was the Mirage G8A, or F.8, a fixed-wing twin-engine fighter capable of intercepting high-speed, high-altitude intruders. It was probably intended to be a French equivalent of an updated Phantom, or maybe even a Tomcat: the one certain thing was that it was expensive, costing two and a half times more than the Mirage F.1. Dassault, with the knack of producing the right design at the right time, was already working on a simpler and cheaper aircraft, which was adopted when the so-called Super Mirage was cancelled.

The reversion to the tailless delta layout was the result of several factors. The trend was for very manoeuvrable aircraft, and the F-16 had emerged as the fighter to beat in this field, but the requirement to intercept

Dimensions	Mirage 2000C
Length (ft/m)	46.50/14.17
Span (ft/m)	29.50/ 8.99
Height (ft/m)	N/A
Wing area (sq ft/m²)	441/40.98
Aspect ratio	1.97

Weights	
Empty (lb/kg)	16,835/ 7,636
Takeoff (lb/kg)	25,928/11,761
Combat (lb/kg)	22,400/10,161

Power	SNECMA M53
Max (lb st/kN)	21,400/95.0
Mil (lb st/kN)	14,400/64.0

Fuel	
Internal (lb/kg)	7,055/3,200
External (lb/kg)	8,758/3,973
Fraction	0.27

Loadings	
Max thrust	0.83 − 0.96
Mil thrust	0.56 − 0.64
Wing takeoff (lb/sq ft/kg/m²)	59/287
Wing combat (lb/sq ft/kg/m²)	51/248

Performance	
Vmax hi	M = 2.35
Vmax lo	M = 1.20
Ceiling (ft/m)	60,000/18,300
Initial climb (ft/min/m/sec)	49,212/250
Takeoff roll (ft/m)	N/A
Landing roll (ft/m)	1,200/410

First flight	Mar 1978

high and fast intruders remained. Analysis led the French to believe that transient performance was more important than sustained turning ability, while advances in aero-dynamics and avionics could overcome the inherent shortcomings of the delta configuration. On the other hand, the tailless delta was well suited to the high-speed, high-altitude mission.

The first step was to utilise relaxed stability combined with quadruplex fly-by-wire to overcome the natural stability of the large delta wing (no manual reversion was provided), accompanied by variable camber: full-

Below: New technology allowed Dassault to revert to a delta planform for the Mirage 2000, producing a remarkable fighter.

span leading edge slats operate automatically at angles of attack greater than five or six degrees when the gear is up, and there are two-section elevons on the entire trailing edge. Small strakes are fitted to the intake sides to reduce the download and produce a vortex, and to offset the moderate power available composites have been used to reduce weight.

In close combat the Mirage 2000 is formidable. Fly-by-wire gives care-free handling, and while the normal stress limits are +9g and −3g the pilot can override them to pull 13.5g in emergency. Doing so produces a high AoA with a consequent drag increase, causing a sharp loss of speed combined with a tighter turn to force a flythrough. The engine would be left at full throttle during this man-oeuvre in order to regain the lost energy quickly. The aircraft is snappy in pitch and acceleration into the roll is fast, the maximum roll rate being 270°/sec. Instantaneous turn is high, according to Dassault between 20° and 30°/sec, while sustained turn at Mach 0.9 and 20,000ft (6,000m) is about 11°/sec. Controllability is good down to 40kt (74km/h) in- ▶

Below: Interesting points seen here are the small strakes on the intakes and the trailing-edge wing/fuselage fillet.

▶ dicated, while at the other extreme brake release to Mach 2 at 49,000ft (15,000m) takes just four minutes.

The radar has been criticised in certain quarters for lack of capability. It is a Thomson-CSF pulse-Doppler multi-mode set, with a maximum detection range of 55nm (100km). The normal weapons fit is two Matra Super 530D medium-range missiles, with a snap-up and snap-down capability, plus two R.550 Magic 2 dogfight missiles. The Super 530D is intended to be able to deal with Mach 3 targets at 80,000ft (24,400m) from a launch altitude of 50,000ft (15,250m), while the heat-homing Magic is reported to pull 50g in the turn. Two DEFA 554 30mm cannon

(rate of fire 1,800 rounds per minute) are also carried internally in the lower fuselage.

It seems that Dassault have done it again, and produced a small, cheap, but very capable fighter, which is already doing well in the export market, and is likely to do better in future.

Armament
Two 30mm DEFA 554 cannon with 125 rounds per gun, two Matra Super 530D and two Matra R.550 Magic missiles

Users
Egypt, France, Greece, India, Peru, United Arab Emirates

Above: The automatic leading edge manoeuvre slats and the two-section elevons are used to produce variable camber.

Below: As with most fast jets, a two-seat conversion trainer is necessary, the Mirage 2000B, seen here with IFR probe.

British Aerospace Sea Harrier

Type: Single-seat carrier-borne counter-air fighter with reconnaissance and strike capability.

The land-based Harrier attack aircraft has served with the Royal Air Force and US Marine Corps for many years, and the first experiments involving ship-board basing and operations date back to February 1963, when test pilots Bill Bedford and Hugh Merewether carried out a series of

Dimensions	Sea Harrier
Length (ft/m)	47.58/14.50
Span (ft/m)	25.25/ 7.70
Height (ft/m)	12.16/ 3.71
Wing area (sq ft/m²)	201/18.69
Aspect ratio	3.17

Weights	
Empty (lb/kg)	13,000/5,900
Takeoff (lb/kg)	19,650/8,915
Combat (lb/kg)	17,125/7,770

Power	Pegasus 104
Max (lb st/kN)	N/A
Mil (lb st/kN)	21,500/95.5

Fuel	
Internal (lb/kg)	5,050/2,290
External (lb/kg)	2,966/1,345
Fraction	0.26

Loadings	
Max thrust	N/A
Mil thrust	1.09 − 1.26
Wing takeoff (lb/sq ft/kg/m²)	98/477
Wing combat (lb/sq ft/kg/m²)	85/416

Performance	
Vmax hi	M = 0.97
Vmax lo	M = 1.25
Ceiling (ft/m)	51,200/15,600
Initial climb (ft/min/m/sec)	50,000/254
Takeoff roll (ft/m)	N/A
Landing roll (ft/m)	N/A

First flight	Aug 1978

takeoffs and landings on the deck of HMS *Ark Royal*. An admiral watching the proceedings later commented that the thing that struck him most was 'the almost complete absence of fright'. Previous jet fighters had been heavier and faster than their predecessors and carrier operations had seemed to become increasingly more perilous. The pilot's comment was that it was easier to stop and then land on a carrier than to land and then stop.

Within a few years British fixed-wing carrier aviation had disappeared, leaving the British fleet to be protected by land-based air power. This was quickly shown to be inadequate, but the American solution — giant aircraft carriers embarking nearly 100 aircraft of which nearly half were dedicated to protecting the carrier and the other ships of the task force — was clearly not affordable.

The British requirement was, in any case, different from that of the Americans, primarily involving the protection of shipping in the North Atlantic sea lanes, far from RAF land bases, where the main air threat was considered to be long-range air-launched missiles fired from Bear aircraft, or submarine-launched missiles using information supplied from Bears via data link. The Bears would be at medium or high altitudes where they could be detected by ships' radar; the ships could then vector a Sea Harrier out to intercept. The Sea Harrier force would be numerically small and was to be carried on the so called through-deck cruisers of the new Invincible class. The Sea Harrier was considered adequate for this job, which was essentially a back-up to land-based air power.

The Sea Harrier was derived directly from the attack Harrier, navalised to minimise corrosion from sea spray and with the cockpit raised to give extra space under the floor and extra side console area. A spin-off of this was that the all-round view was much improved, especially rearward. The avionics were optimised for the task, ▶

▶ and included the Blue Fox pulse radar, which had no real look-down capability but was adequate for the medium and high altitude work envisaged. The two 30mm Aden cannon pods were retained, and provision was made for two Sidewinder missiles.

It is often claimed that the Argentinian forces in the 1982 South Atlantic War were disadvantaged by the distance from their bases over which they were forced to operate, but the disadvantages suffered by the Sea Harrier force are rarely pointed out. The two British carriers, *Hermes* and *Invincible,* had to be kept well back out of harm's way, as the loss of either could have been disastrous, which increased the distance to the patrol areas and reduced the time on patrol.

With no airborne early warning and ship-borne control in many cases minimal, the Sea Harriers were forced to resort to the wasteful practice of standing patrols, which meant that they were often outnumbered, albeit in relatively small engagements. With little radar look-down capability, most sightings were made visually, and many chances were missed.

Finally, too much has been made of the AIM-9L Sidewinder's all-aspect capability. In practice, it was rarely if ever used, and even though most missile launches were made from astern, the kill ratio achieved was only about 67 per cent. At very low level, where the majority of engagements took place, the brochure range of AIM-9L is severely curtailed; accounts of launches at co-speed Skyhawks from astern at visual range (Skyhawk is small, and visual range short) refer to them falling short. All in all, it is quite remarkable that the Sea Harriers did

as well as they did. For no losses in air-to-air combat, they downed two Mirage IIIs, nine Daggers, eight Skyhawks — all but two Skyhawks with Sidewinders — plus a Pucara, a Canberra, and a Hercules.

In close combat, the Sea Harrier is excellent. It is small, with a smokeless engine, and therefore difficult to see. It is an odd shape, and it can be difficult to tell what it is doing from some aspects. The STOVL characteristics hardly need comment, and from a short rolling takeoff it can reach 40,000ft (12,000m) in just two minutes. It turns well (and even better if VIFF is employed), and is snappy in the rolling plane. The modest fuel fraction is deceptive, as

the lack of augmentation and the economical turbofan engine enable it to go farther than one might expect. Maximum interception radius is stated to be 400nm (740km).

Particularly spectacular is its acceleration from low speeds, which is probably better than any other fighter in the world. This is the result of the high thrust loading in a regime where afterburning is inefficient. All else being equal, in combat with a conventional afterburning fighter, it will be the non-STOVL type that breaks off first for shortage of fuel.

Much has been made of the Sea Harrier's response to vectoring of the engine thrust in forward flight (VIFF). While the technique certainly allows the Sea Harrier to pull some otherwise impossible manoeuvres, it was not used in the South Atlantic: it would normally be used only in exceptional circumstances, as it tends ▶

Below: A Sea Harrier lands aboard HMS *Illustrious* off the Falklands, where the type proved invaluable in the 1982 war.

▶ to be somewhat imprecise in control — to force an overshoot in an emergency, or in small amounts to nibble off a few degrees of angle to bring the guns to bear.

The Sea Harriers of the Royal Navy will shortly undergo upgrading to FRS.2 standard. It was found in 1982 that just two Sidewinders was inadequate, and made for short combat persistence, and RN Sea Harriers already carry four. The update involves a pulse-Doppler radar, the Ferranti Blue Vixen, which will give a true look-down/shoot-down capability, and in order to make the most of the new radar the weapons fit will be increased to four AIM-120 Amraams, Sea Harrier being the first European fighter scheduled to carry the new missile, which should be much more satisfactory should it become necessary to take on Soviet Backfire and Blackjack bombers.

India flies the Sea Harrier FRS.51, which varies from the FRS.1 mainly in details of the avionics and in the fact that its main weapon is the Matra R.550 Magic. Italy is expected to order Sea Harrier shortly to equip its new helicopter carrier *Giuseppe Garibaldi,* which has a ski-jump on the flight deck.

Various schemes are afoot for future Sea Harriers. One of these is SCADS (Shipborne Containerized Air Defence System) which will allow quite small ships to carry their own fighter. Another obvious scheme is for a supersonic Sea Harrier, using plenum chamber burning. This project, known as ASTOL (Advanced STOL) is likely to produce a rather larger aircraft with more advanced avionics, capable of maintaining combat air patrol for two and a half hours at an unspecified distance from the mother ship.

If ASTOL proceeds, it will give the Harrier family a credibility that it has lacked until now. Almost no advanced nation can see much virtue in a subsonic fighter, and certainly one of the most telling arguments against the Sea Harrier in recent times has been 'how do you intercept a supersonic Backfire with a subsonic Sea Harrier'. A supersonic STOVL aircraft might just prove to be the right product at the right time in the world marketplace.

Armament
Two 30mm Aden cannon with 100 rounds per gun, two or four AIM-9L Sidewinder or R.550 Magic missiles (later four AIM-120 Amraam)

Users
India, UK.

Above: One of the Indian Navy's Sea Harrier FRS.51s shows its Matra R.550 Magic missile armament during a pre-delivery flight over southern England.

Below: The Sea Harrier FRS.2 will be the first European fighter to be armed with Amraam missiles; new avionics and a new radar will also be installed.

Dassault-Breguet Rafale

Type: Single-seat multi-role fighter.

Regardless of the fact that Rafale is a proposed replacement for the Jaguar, Mirage IIIE and Mirage 5 from 1995, it seems obvious that Dassault-Breguet have done everything possible to produce a close combat fighter to beat all others. Rafale combines the delta configuration with relaxed stability and fly-by-wire, or maybe even fly-by-light in the future; it has movable canard foreplanes to reduce the downloads and to create a vortex across the wings; and the fact that the foreplanes are movable allows direct lift to be used, increasing the pointability of the fighter. It has, by modern standards, very low wing loading, and extremely high thrust loading — better, in fact, in military thrust than many other fighters have using afterburning — and at the 1986 Farnborough air show Rafale A demonstrated a sustained turn rate of 24°/sec at low altitude, an exceptional performance. Six missiles plus a 30mm cannon give good combat persistence, as does the high, although not excessive, fuel fraction.

In common with the F-16, the fighter that has for many years been used as a yardstick for close combat, Rafale has an inclined seat, a sidestick controller and a teardrop canopy giving 360° vision. In-flight

Dimensions	Rafale A	Rafale B
Length (ft/m)	51.84/15.80	N/A
Span (ft/m)	36.08/11.00	N/A
Height (ft/m)	16.40/ 4.00	N/A
Wing area (sq ft/m²)	506/47.00	474/44.00
Aspect ratio	2.57	N/A
Weights		
Empty (lb/kg)	20,950/ 9,500	18,750/ 8,500
Takeoff (lb/kg)	30,870/14,000	28,670/13,000
Combat (lb/kg)	26,450/12,000	24,250/11,000
Power	2xF404	2xSNECMA M8
Max (lb st/kN)	16,000/71.1	16,800/75.0
Mil (lb st/kN)	10,600/47.1	11,500/51.0
Fuel		
Internal (lb/kg)	8,800/4,000	8,800/4,000
External (lb/kg)	N/A	N/A
Fraction	0.29	0.31
Loadings		
Max thrust	1.04 — 1.21	1.17 — 1.39
Mil thrust	0.69 — 0.80	0.80 — 0.95
Wing takeoff (lb/sq ft/kg/m²)	61/298	61/295
Wing combat (lb/sq ft/kg/m²)	52/255	51/250
Performance		
Vmax hi	M = 2	M = 2
Vmax lo	M = 1.2	M = 1.2
Ceiling (ft/m)	1,300/400	1,300/400
Initial climb (ft/min/m/sec)	980/300	980/300
Takeoff roll (ft/m)	N/A	N/A
Landing roll (ft/m)	N/A	N/A
First flight	Jul 1986	N/A

Rafale B weights, fuel fraction and loadings estimated

refuelling will be incorporated, and the ability to operate from a 500m strip with two AAMs and guns loaded was a basic requirement.

Rafale A is purely a technology demonstrator to prove the concept, and the production Rafale B will be smaller. Rafale was once a contender for the European Fighter Aircraft, but the French dropped out because they wanted a specifically lighter aircraft: Rafale B will be about a tonne lighter than the A; the overall dimensions are to be slightly smaller and the wing area reduced, and extensive use of composites will save weight. In fact, it is the weight saving that shifts Rafale B into an exceptional thrust-loaded class and promises outstanding performance. That bodes well for the export market, as does the light empty weight, which can be roughly

Above: The intake adopted for Rafale compromises between the side and chin positions.

equated with cost; the heavier Eurofighter will be more expensive.

Rafale is to carry a pulse-Doppler multi-mode radar with a maximum search range of 55nm (100km) against fighter-sized targets, and allow a look-down/shoot-down capability. It will be able to track up to eight targets while continuing to search and be able to assess raids and allot priorities to targets. It will carry the Matra MICA missile, a French Amraam equivalent with inertial mid-course guidance and either IR or active radar terminal homing, as well as Magic 2.

A carrier-based version with beefed-up gear and an arrester hook, plus provision for catapult launching, will be some 1,100lb (500kg) heavier. The gear is to be stressed for a sink rate of 13ft (4m) per second, rather low judged by American standards, which are approximately double that. Ski-jump takeoffs have also been mentioned by one source, but this hardly seems likely in a carrier context, as the ski-jump would be an obstruction in the event of a bolter. No order has been placed for Rafale, but development is proceeding, and it certainly seems a very potent fighter.

Armament
One 30mm DEFA 554 cannon, four MICA medium-range missiles, two R.550 Magic 2 missiles

Below: Seen here is Rafale A, a technology demonstrator for the proposed Rafale B.

Eurofighter EFA

Type: Single-seat counter-air fighter with interception and attack as secondary capabilities.

The origins of the Eurofighter lie in the requirements of three nations, France, Germany and the United Kingdom. France has dropped out and gone ahead with Rafale independently while Italy and Spain have joined Germany and the UK in a consortium to build the new fighter for Europe. The Eurofighter is bigger and rather heavier than Rafale, the factor which led to France abandoning the joint project.

The design has been influenced by current weaponry advances, and started from the basic premise that most future combats would begin beyond visual range using a launch-and-leave weapon such as Amraam, after which the range would swiftly close down to visual. The first need was to accelerate hard to impart the maximum energy to the missile on launch and so increase its manoeuvrability and probability of kill at near-maximum range; after launch heavy manoeuvring would be called for to evade a hostile missile attack which it was presumed would be launched at the same time.

In both phases Eurofighter will benefit from a thrust:weight ratio which is high by any standards. The airframe is optimised for high agility

Dimensions	BAe EAP
Length (ft/m)	57.50/17.53
Span (ft/m)	36.65/11.17
Height (ft/m)	18.13/ 5.53
Wing area (sq ft/m²)	538/50.00
Aspect ratio	2.5

Weights	
Empty (lb/kg)	22,050/10,000
Takeoff (lb/kg)	34,000/15,420
Combat (lb/kg)	29,250/13,270

Power	2xRB199 Mk 104D
Max (lb st/kN)	16,500/73.3
Mil (lb st/kN)	N/A

Fuel	
Internal (lb/kg)	9,500/4,310
External (lb/kg)	N/A
Fraction	0.28

Loadings	
Max thrust	0.97 — 1.13
Mil thrust	N/A
Wing takeoff (lb/sq ft/kg/m²)	63/308
Wing combat (lb/sq ft/kg/m²)	54/265

Performance	
Vmax hi	N/A
Vmax lo	N/A
Ceiling (ft/m)	N/A
Initial climb (ft/min/m/sec)	N/A
Takeoff roll (ft/m)	N/A
Landing roll (ft/m)	N/A

First flight	Aug 1986

and BAe have predicted that it will outmanoeuvre any aircraft now flying. The main emphasis is on instantaneous turn, with the high thrust loading replacing lost energy quickly. The wing is a kinked delta which gives the best combination of low wave drag and aspect ratio, with the steeply swept inboard generating vortices to increase lift at high AoA. Automatic wing camber is also featured as a function of velocity and AoA, and to minimise supersonic drag, while the lower lip of the intake is hinged to optimise airflow at high AoA.

Right: The EFA in mock-up form representing the configuration finalised in the Spring of 1986. The apparent curvature of the wing is an optical illusion.

Armament
One 25mm Aden or 27mm Mauser cannon; four AIM-120 Amraam or Sky Flash missiles, two Asraam or AIM-9L Sidewinder missiles

Users
Germany, Italy, Spain, UK

Left: The EFA mock-up, on show at Farnborough 86, showing the all-moving canard surfaces and the box-section nodding type intake.

Below: An artist's impression of EFA in flight, showing the basic configuration, but with a fin that is far too small.

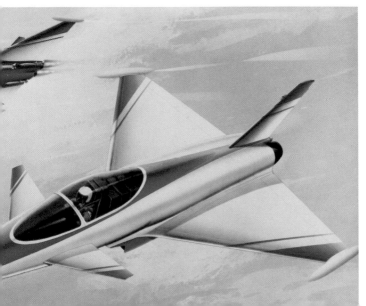

Saab JAS 39 Gripen

Type: Single-seat multi-role fighter for the counter-air attack and reconnaissance missions. A two-seat fully combat-capable trainer may be built.

Much of the tabular material on the Gripen is of a highly speculative nature: the only figure released for the weight is that it is half that of the Viggen, and the other figures given are based on that statement; any further calculations would simply compound the errors.

Gripen is the lightest of the new generation of unstable canard delta fighters by quite a margin. While it is definitely a state-of-the-art fighter, with advanced avionics and considerable use of composites, it is believed that its performance will not exceed that of the F-16, although it might feature direct lift sideforce control, which the Swedes are known to favour for close combat. First flight is scheduled to take place in 1987.

Armament
One 27mm Mauser BK27 cannon, four AAMs.

User
Sweden

Dimensions	JAS 34 Gripen
Length (ft/m)	45.90/14.00
Span (ft/m)	26.25/ 8.00
Weights	
Empty (lb/kg)	14,000/6,350
Takeoff (lb/kg)	18,000/8,170
Combat (lb/kg)	15,500/7,030
Power	
Max (lb st/kN)	18,000/80
Mil (lb st/kN)	12,140/54

Right: The Gripen cockpit is clearly state of the art, with three multi-function CRT displays and a wide-angle HUD.

Below right: Gripen follows the current fashion for delta wings with canard foreplanes for greater manoeuvrability.

Below: A true multi-role fighter, Gripen is shown equipped for air combat (top) and with air-to-ground weapons.

Combat Tactics

Above: Small is beautiful in close combat — a Top Gun Skyhawk breaks away after a successful attack on a Tomcat.

Below: Phantom versus Tomcat. Grumman Chief Test Pilot Chuck Sewell easily outmanoeuvres US Navy test pilot Lt D. Walker.

The modern fighter combines detection and countermeasures, aerodynamics and power, along with many different types of weapons, to form an integrated weapons system. Its capabilities can be considerably increased by such force multipliers as airborne early warning and control, dedicated countermeasures aircraft and the humble tanker which, as a flying gas station, can extend a fighter's range and endurance considerably. Capabilities vary considerably from type to type, but it should always be remembered that the most advanced fighter is badly disadvantaged by the most basic if the latter is at six o'clock and within firing parameters.

In the preceding sections we have examined the detection systems and the weaponry in general terms and the aircraft themselves in detail. There is, however, one weakness that has not been touched, one which is so universal and so obvious that it is seldom seen as a weakness, namely that aircraft weapons can only be fired forward.

The more basic fighter/weapon combinations require the nose of the fighter to be pointing directly at the target in what is known as boresight mode while the more sophisticated combinations have a certain amount of off-boresight capability. The latter is often limited by the aiming system, which is shown on the HUD, and which has led to the development of wide angle HUDs. The next step will be to use helmet-mounted sights, with the aim information projected onto the pilot's visor.

In air combat terms, aiming or weapon system limitations mean that with simple systems the nose of the fighter must be pointed at or just ahead (for lead) of the target, while the more advanced systems demand that the nose is aligned with the general direction of the target.

The capability of the weapon/aiming system determines the degree to which the fighter must point toward the target; it also determines the area that the potential target should keep out of if possible, to deny the attacker a shot. Against an opponent using a missile such as the Matra Super 530D, which is designed to snap up and destroy a target flying up to 30,000ft (9,150m) above the launch aircraft this has little relevance, but it has a great deal of relevance to close combat at visual ranges: while it may well prove impossible for the target

Cone of vulnerability

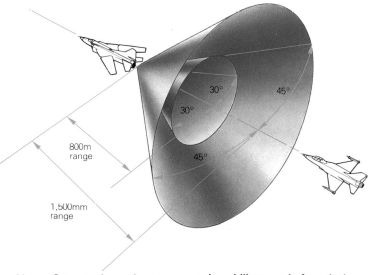

Above: Gun attacks are best made from astern, with the attacker taking a position in the vulnerability cone before closing to a lethal position within 30° angle off.

Direct pitch control

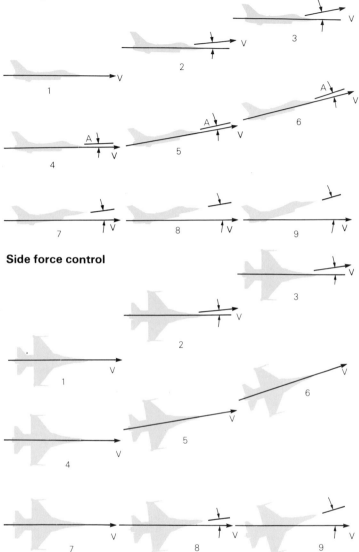

Side force control

Top: Weapons aiming is primarily a matter of getting the fighter's nose on target. Figures 1,2,3 show the fighter climbing while maintaining a level attitude in what is known as vertical translation mode. Direct lift at a constant angle of attack (4,5,6) gives precision control in the vertical plane, while pitch pointing (7,8,9) involves maintaining the original course and speed.

Above: Side force control confers similar benefits to pitch mode control. In lateral translation (1,2,3), the fighter drifts sideways while maintaining its original course; direct sideforce control (3,4,5) permits target tracking, and in yaw pointing (7,8,9) the heading is changed while the flightpath remains constant, improving aim control. (A = angle of attack, V = flight path).

fighter to keep out of the danger area all the time, it is helpful to make the opponent's missile tracking system work as hard as possible.

If the launching fighter can be made to manoeuvre hard at the same time there will be a further increase in possiblity of the shot failing, as there is a tendency for a missile to weather-cock into the relative wind, which is the direction of the airflow past the fighter, regardless of which way it is pointing, and at high AoA, the two can differ by 20° or more. Pointability is, therefore, an asset in a fighter, which explains the current preoc-cupation with direct lift and direct side-force control. But the nose of the fighter still has to be bought to bear, more or less, on the target, and for gunshots boresight aiming is an absolute requirement for the sighting system.

To get the best out of his expensive hardware, the fighter pilot is taught basic manoeuvres which he develops and varies with experience. Some of the techniques concern BVR com-bat, which is to a large degree a ques-tion of getting the best out of the detection and ECM systems and which has been covered earlier; others concern teamwork, tactical formations, and air combat man-oeuvres.

Teamwork is essentially a matter of mutual cover and backup, increasing both the offensive and defensive capability of the formation as a whole. It includes tactical doctrines such as Loose Deuce, Double Attack and others, a discussion of which is unfortunately precluded by space limitations.

Formations are variable, and are determined by the number of fighters involved, the nature of the mission and the nature of the opposition. The basic element is always the pair, who provide mutual cross-cover and try hard, but not always successfully, to stay together. The pair normally flies

Search and reporting

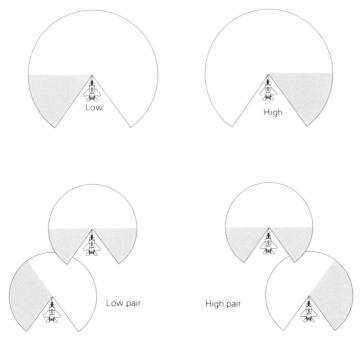

Above: The primary (unshaded) and secondary areas of responsibility for a pair (top) and a four-ship formation. Three- **quarters of a pilot's time is spent searching his primary direction — high and low. near and far — and all sightings are reported.**

Inward turnabout

High man goes low

Above: This cross-turn is used to allow an element of two aircraft to reverse course with no spatial displacement. If altitudes are staggered, the high man goes low and the low man high, checking each other's six as they pass.

Low man goes high

The sandwich

Attacker threatened by other defender

Defender turns hard

Below: A pair flying in wide abreast formation can sandwich an attacker. The defending fighter turns outward while his wingman turns inwards.

a formation called combat spread, which is a wide line abreast. If the tactical situation permits, the fighter nearest the line of the sun flies stepped down, so that the vertical angle of the sun is different for each fighter, and there is less chance of a surprise attack succeeding from this direction.

Lateral spacing varies. The idea is that if one fighter is attacked, the other can come to its assistance with the least delay, and ideally this means one fighter turn diameter at the height and speed selected. In conditions of poor visibility the distance would have to be reduced to about half that over which a visual contact could be made. If a radical change in direction is called for the aircraft cross over in the turn, while for a course reversal an inward turnabout is favoured, with the high man going low and the low man high. This has

Attackers

the double advantage that there is no lateral displacement of the patrol line, if patrol line it is, while in the turn, the pilots are well placed to check each other's blind spots below and astern.

Many other permutations are possible, usually but not always in combinations of two. Two pairs abreast and stepped down towards the sun is common, while pairs in trail can have advantages. Finally, if a four-ship formation is scheduled and one of the aircraft is found to be unserviceable at the last moment, a three-ship layout with the pair trailing the leader could be used.

Echelon and trail have been used for pairs, but as the rear machine has no cover they are tactically unsound. They can be used in a controlled environment, such as in an engagement where no unexpected hostile fighters are likely to appear, either to visually identify bogeys, when the leader makes the identification and the wingman locks up his radar for a BVR shot, or for the leader to 'drag' the bogeys and set them up for a rear quarter shot by the wingman. In an uncontrolled environment, or against fighters believed to have BVR or all-aspect weapons, such practices are gradually becoming too risky; on the other hand, combat spread is so popular that it may make the pair too predictable.

Air combat manoeuvres, or basic fighter manoeuvres as they are sometimes called, are essentially methods of pointing the fighter's nose at the target while preventing him from reciprocating.

The manoeuvre phase of combat commences at the moment that the attacker loses the advantage of surprise. Of course, it is possible for

Defensive split

Below: The defensive split is used to force two attackers to concentrate on one aircraft, leaving the other free. One defender pulls high while the other goes low; the defenders must then make every effort to rejoin for mutual support.

Defenders

Air combat phases

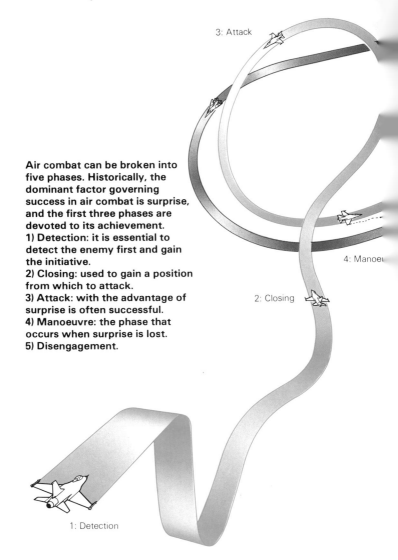

Air combat can be broken into five phases. Historically, the dominant factor governing success in air combat is surprise, and the first three phases are devoted to its achievement.
1) Detection: it is essential to detect the enemy first and gain the initiative.
2) Closing: used to gain a position from which to attack.
3) Attack: with the advantage of surprise is often successful.
4) Manoeuvre: the phase that occurs when surprise is lost.
5) Disengagement.

3: Attack

4: Manoeu

2: Closing

1: Detection

close combat to start from a neutral position, with the opponents spotting each other simultaneously, in relative positions which give neither the advantage. Defeating missiles that have already been launched has been covered earlier: manoeuvre combat is about fighter defeating fighter.

The usual direction of a close-range surprise attack is from the astern quarter. The attacker will be trying for a heart-of-the-envelope heat missile shot, or a tracking gun shot. In the first case he will be trying to drop in exactly astern outside minimum missile range, while for a gun shot he will be pulling lead on the target, with his nose pointed slightly ahead, while rolling to get in plane with the motion of the target. The defender's first move is the break.

The break is a maximum rate turn into the direction of the attack. If a

5: Disengagement

missile attack is expected, the aim will be to create as much angle-off as possible in the time available, while if the situation looks like a gun attack, the break must be combined with a roll or pull out of the plane of motion; otherwise the attacker is presented with a snap-shot at a planform target, the biggest possible presented area.

Often the attacker closing for a stern shot will be travelling faster than the defender. If the speed disparity is too great the attacker will overshoot, and the defender should be alert to the possibility of reversing his turn to gain the advantage. This will become apparent as the attacker goes wide in the turn; if his relative speed takes him wide quickly an early reverse is called for, but if he drifts slowly the reverse should be made late or not at all. If the attacker does not overshoot, either from not having excess speed or simply through being better in the turn, the defender must

The break

Right: The break is a maximum rate turn designed to spoil an opponent's aiming solution and is often combined with an out-of-plane manoeuvre. It may force an attacker into an overshoot, when the defender can reverse to turn the tables.

Defender turns sharply into direction of attack

Attacker about to achieve firing position

stay in the hard turn and try to increase, or at least maintain, the angle-off already generated. Against a better turning adversary this may not work: it becomes a question of holding him off until help arrives, or until he runs low on fuel.

Altitude permitting, the spiral dive is the next move for the defender, trading altitude for energy to maintain a hard turn. This can sometimes be accompanied by a reduction in power while flattening out the dive, which is difficult for the attacker to spot, and may result in an overshoot, or at least a neutral position.

If all else fails, the last resort is jinking — a series of turns, skids, yaws, and pitch-ups intended to produce an unpredictable flight path in order to spoil the attacker's aim. At very low levels jinking can be highly effective, as the countryside at close range can quite distract the attacker from his aim.

If an overshoot is forced at the outset, and the defender reverses his turn, a scissors may occur. This is a series of turn reversals aimed at making an opponent fly through and end up in front, where he in turn becomes the quarry. Scissoring with a more

Overshoot in spiral dive

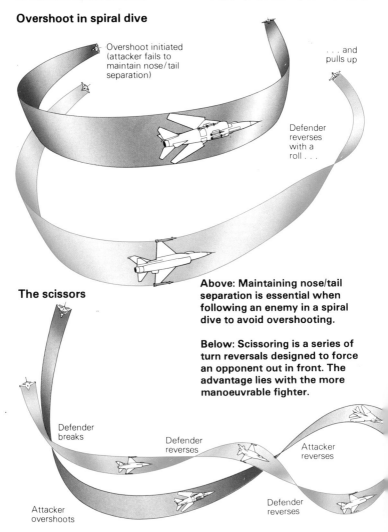

Overshoot initiated (attacker fails to maintain nose/tail separation)

. . . and pulls up

Defender reverses with a roll . . .

The scissors

Above: Maintaining nose/tail separation is essential when following an enemy in a spiral dive to avoid overshooting.

Below: Scissoring is a series of turn reversals designed to force an opponent out in front. The advantage lies with the more manoeuvrable fighter.

Defender breaks

Defender reverses

Attacker reverses

Attacker overshoots

Defender reverses

The Split S

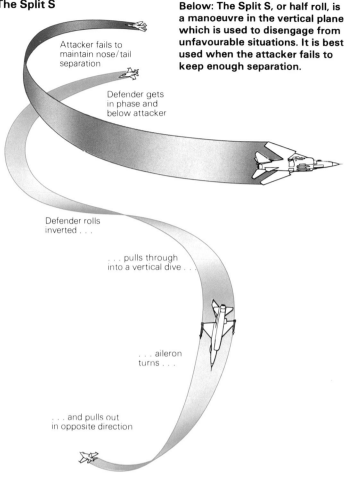

Below: The Split S, or half roll, is a manoeuvre in the vertical plane which is used to disengage from unfavourable situations. It is best used when the attacker fails to keep enough separation.

Attacker fails to maintain nose/tail separation

Defender gets in phase and below attacker

Defender rolls inverted . . .

. . . pulls through into a vertical dive . . .

. . . aileron turns . . .

. . . and pulls out in opposite direction

Attacker forced out in front

manoeuvrable opponent is not recommended, since acceleration into the turn gives a considerable advantage, as does instantaneous turn rate.

The vertical rolling scissors is a variation on a theme. It is usually carried out in a steep climb or dive, with complete barrel rolls replacing the turn reversals. In a protracted ascending vertical rolling scissors the advantage will lie with the fighter which has the best sustained climb rate. Disengagement from a vertical rolling scissors is best made with a Split S and a lot of hope.

The Split S is a disengagement manoeuvre which can be used from many positions, provided only that enough altitude is available to allow the pull-out. It consists of a rapid roll to an inverted position followed by a vertical dive, aileron turning on the way down ready for a pull-out in the desired direction, usually the opposite course to that of the opponent. Ideally it should be used from a position of neutrality, following an overshoot, or after a head-on pass, or from a position where the opponent has lost sight for a moment. It presents a difficult missile target to an attacker because of the ground clutter, while one of the time-honoured rules of air combat states 'never follow anyone down'. A further advantage is that, like most three-dimensional or vertical manoeuvres, it allows a fighter to turn 'square corners' relative to the horizontal plane.

The High-g Barrel Roll is another three-dimensional manoeuvre, one which can be used against an attacker closing fast from dead astern, It involves a speed loss — up to 100kt (185km/hr) in some cases — while the spiral path slows the forward velocity vector. Against a much faster attacker it usually results in a fly-through, but a co-speed or only slightly faster attacker will follow the defender through the move and end within easy gun range. The manoeuvre can also be used to defeat a head-on missile attack by causing large angular changes in the seeker head tracking rate.

Often a very fast attacker will be unable to stay with the target fighter in a hard turn: the solution here is the High YoYo, which uses the vertical to prevent overshooting. Relaxing his

The High-g Barrel Roll

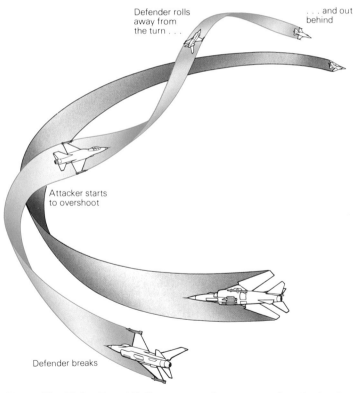

Defender rolls away from the turn . . .

. . . and out behind

Attacker starts to overshoot

Defender breaks

Above: The High-g Barrel Roll can be used to defeat either a fast attack from astern or a head-on missile shot. In the former case it can cause the attacker to overshoot; in the latter it compounds missile tracking problems.

turn a little, the attacker pulls high and rolls inverted both to retain visual contact and to avoid excessive negative g. Speed loss in the pull-up and climb, combined with the extra g of gravity, reduces his turn radius as he comes over the top, leaving him well placed to come down into the target's stern quarter. The High YoYo is very difficult to judge correctly, and if the nose is pulled too high it allows the defender to disengage with a Split S. If the tactical situation permits, a series of small High YoYos is often preferable.

The Rollaway, or Vector Roll, is a variation on the High YoYo used in much the same circumstances; the attacker relaxes his turn and pulls up, then rolls away from the direction of

The High YoYo

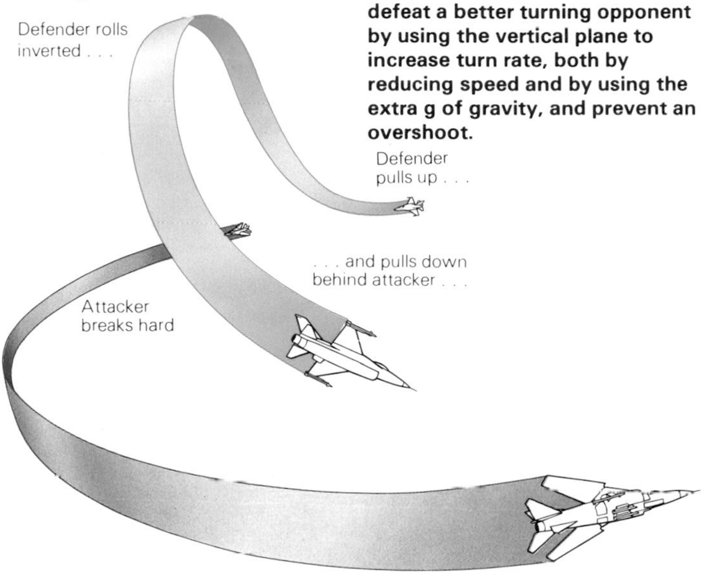

Defender rolls inverted . . .

Below: The High YoYo is used to defeat a better turning opponent by using the vertical plane to increase turn rate, both by reducing speed and by using the extra g of gravity, and prevent an overshoot.

Defender pulls up . . .

. . . and pulls down behind attacker . . .

Attacker breaks hard

The Rollaway

. . . rolls away from direction of turn . . .

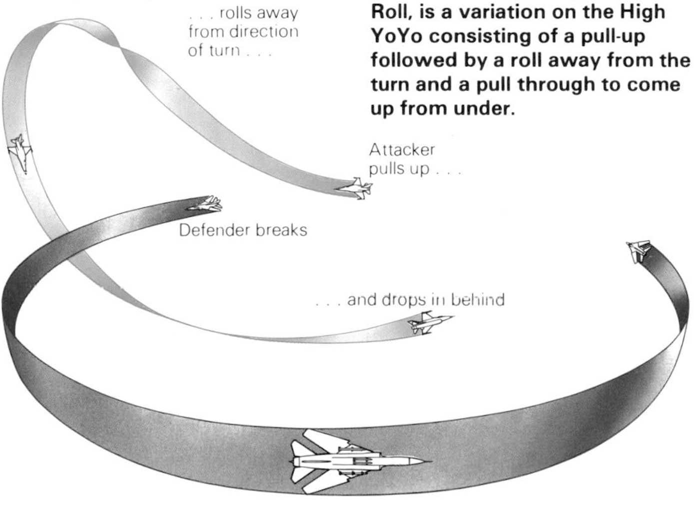

Below: The Rollaway, or Vector Roll, is a variation on the High YoYo consisting of a pull-up followed by a roll away from the turn and a pull through to come up from under.

Attacker pulls up . . .

Defender breaks

. . . and drops in behind

the turn, pulling over the top then down and across the circle into the rear hemisphere of the target.

The opposite case to the High YoYo is the Low YoYo, which is used to catch a faster, hard turning bogey, or to break a stalemate turning situation. It is based on trading height for speed, and is executed by dropping the nose on the inside of the turn and gaining speed in a shallow dive across the circle before pulling back up in the target's stern quarter.

The main problems facing attacking fighters are to do with achieving the best possible firing solution while maintaining the element of surprise. Surprise demands that the firing position be achieved quickly — the longer the delay, the more chance the target has of spotting the attacker — and the problems are positional and dynamic. Rarely is a target detected at an ideal angle to be attacked, and if there is no speed disparity at the detection stage, there will be as soon as the attacker accelerates to try and reach a firing solution. The problems arise from excessive angle off and excessive speed, both of which are major contributors to an overshoot: the manoeuvres to counter an overshoot in either case have been given, but there remains one final counter to an overshoot caused purely by speed.

Lag Pursuit consists of avoiding overshooting by taking up a position on the outside of the target's turn and matching his turn rate, at the same time maintaining a speed (and therefore energy) advantage while turning on the wider radius. This, incidentally, is the one fighter manoeuvre where sustained turn radius really counts for the defender.

The real problem for the defender is that the attacker is out of sight, below and astern, and he cannot see what is happening. Unless each fighter has a free wingman, who can keep them informed of the situation, Lag Pursuit rapidly develops into a game of chicken. The attacker is preoccupied with watching his oppo-

Lag Pursuit

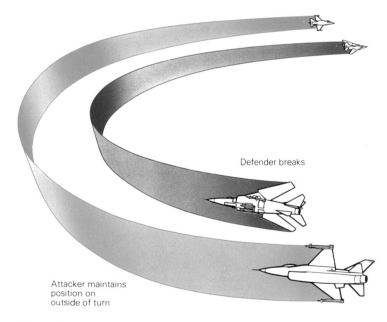

Defender breaks

Attacker maintains
position on
outside of turn

Above: Lag Pursuit is another method of avoiding overshoots. When the defender breaks the attacker takes up position on the outside of the turn and matches the defender's rate of turn, maintaining a position on the defender's belly side from which he cannot be seen, thus keeping the initiative.

Above: The pursuing Tomcat is in the saddle for a guns shot against the USN Phantom.

The Low YoYo

. . . and pulls up within firing range

Attacker gains speed in shallow dive . . .

Above: The Low YoYo can be used to close the range on a co-speed target in straight pursuit by trading altitude for speed in a shallow dive, then pulling up to attack when weapons range is reached. The best counter to this is for the defender to keep sight by weaving, and match the dive.

The Barrell Roll Attack

Defender turns to force overshoot
but still comes under attack

**Left: The Barrel Roll Attack
prevents an overshoot by using a
three-dimensional manoeuvre to
change the angle off.**

The Vertical Reverse

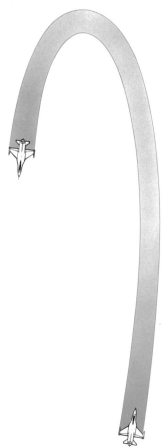

nent for the next move, and will very
quickly become concerned about
what is going on around him, and
worried about catching a belly missile
from a unseen opponent, while the
defender, not knowing what is going
on behind him, has the constant tem-
pation to reverse in order to find out.
That would not be healthy, and the
defender's best move would be to
tighten into a spiral dive.

There are many cases where a
three-dimensional manoeuvre can be
used to counter a two-dimensional
one. Gravity can be used to tighten
turn radii, while vertical manoeuvres,
especially the three-dimensional
variety such as the barrel roll, can be
used to turn what are, in flat plane
terms, square corners.

The Barrel Roll Attack is one example. Similar in nature to the Rollaway,
it is also used to prevent overshooting
a hard turning target: the attacker
rolls wings level then pulls up hard,
rolling away from the direction of turn
before pulling the nose through in the
direction of the target.

The Vertical Reverse is basically a
repositioning manoeuvre carried out
at the top of a climb: as speed falls

**Above: The Vertical Reverse is
the modern equivalent of the
stall turn. As flying speed is lost
the fighter rudders sharply
around, swapping ends very
quickly to confront a pursuer
head-on.**

away the fighter is given full rudder, which drops it over into a vertical dive. Most recent fighters can do it, but it is beyond the capabilities of some of the older types. The Sea Harrier, using vectored thrust, can perform it magnificently and most unexpectedly. It can be used to terminate a vertical ascending scissors.

The Immelmann Turn is another repositioning manoeuvre: a pull-up to the vertical is completed by rolling out on the opposite heading at the top of the climb, or in any desired direction by aileron-turning during the vertical climb.

The foregoing are part of the fighter pilot's standard bag of tricks when engaged within visual range. They are designed to cope with situations that may occur in combat, and any one of them may need to be pulled out of the hat at a moment's notice to meet a given situation. But it

The Immelmann

Above: The Immelmann is used to reposition and change the heading by any angle while remaining over the same spot. Effectively flying a square corner in the horizontal plane, but with an altitude increase.

would be unrealistic to assume that a fighter pilot would engage in an extended dogfight at one versus one, going from one move to the next in a set sequence. A move that works today may not work tomorrow, and when real missiles are in the air, anything more than a break turn tends to feel fancy.

The real task of the fighter pilot is the strike rapidly, using every possible advantage to ensure surprise, carry out a successful attack, then disengage and reassess the situation. In even a small combat such as a two versus two, each pilot has to keep track of three other aircraft and what they are doing, fly his fighter, use the weapon systems, watch the fuel gauge carefully, and still manage to keep a notional look-out for any other bogeys joining the fray. He very quickly overloads, and his situational awareness sinks to zero.

In every air conflict there have been a few pilots who have beaten the odds. They are the aces, and the quality that sets them apart from the rest is their apparent ability to keep track of what was going on around them to a much greater extent than the average.

The electronic wizardry of the modern fighter and its force multiplier systems has made such awareness even more difficult than ever before. There are, for example, certain set pieces for opening moves for

two versus two, and two versus many, but the choice depends on knowing in advance what the opposition is, what its capabilities are, whether it has all-aspect BVR missiles, and so on.

Typical of this genre is the heart attack, or bracket. Detecting two bogeys approaching in combat spread, a pair splits wide to outflank them, turning in just before an abreast position is reached, with each pilot performing a stern conversion on the fighter *furthest* from him. If they achieve surprise, all well and good; the natural sequel is a stern quarter attack, a heat missile launched, and mission accomplished. If they are seen visually as they curve in

behind, each bogey will turn into the attack, because that is what the book says they should do, and in doing so they will present their hot tailpipes obligingly toward their real as opposed to their assumed attackers.

If however, it turns out that they detected first by whatever means, AEW perhaps, and are carrying BVR launch-and-leave missiles, they simply turn toward one or other of the inbound fighters, aquire and launch, then turn back for a two versus one on the survivor. The more complex the situation, the more theoretical set piece plays become. Historically, hit and run tactics have always proved best. There can be little doubt that this is a situation that will continue.

OTHER SUPER-VALUE MILITARY GUIDES IN THIS SERIES......